Pale Rider

BY THE SAME AUTHOR

The Doctor

The Quick

Rue Centrale: Portrait of a European City

Pale Rider

The Spanish Flu of 1918 and How it Changed the World

LAURA SPINNEY

JONATHAN CAPE
LONDON

1 3 5 7 9 10 8 6 4 2

Jonathan Cape, an imprint of Vintage Publishing,
20 Vauxhall Bridge Road,
London SW1V 2SA

Jonathan Cape is part of the Penguin Random House group of companies
whose addresses can be found at global.penguinrandomhouse.com.

First published in the United Kingdom by Jonathan Cape in 2017

penguin.co.uk/vintage

A CIP catalogue record for this book is available from the British Library

ISBN 9781910702376

Typeset in India by Integra Software Services Pvt. Ltd, Pondicherry

Printed and bound in Great Britain by TJ International Ltd, Padstow, Cornwall

Penguin Random House is committed to a sustainable future for
our business, our readers and our planet. This book is made from
Forest Stewardship Council® certified paper.

For RSJF and the lost generations

Contents

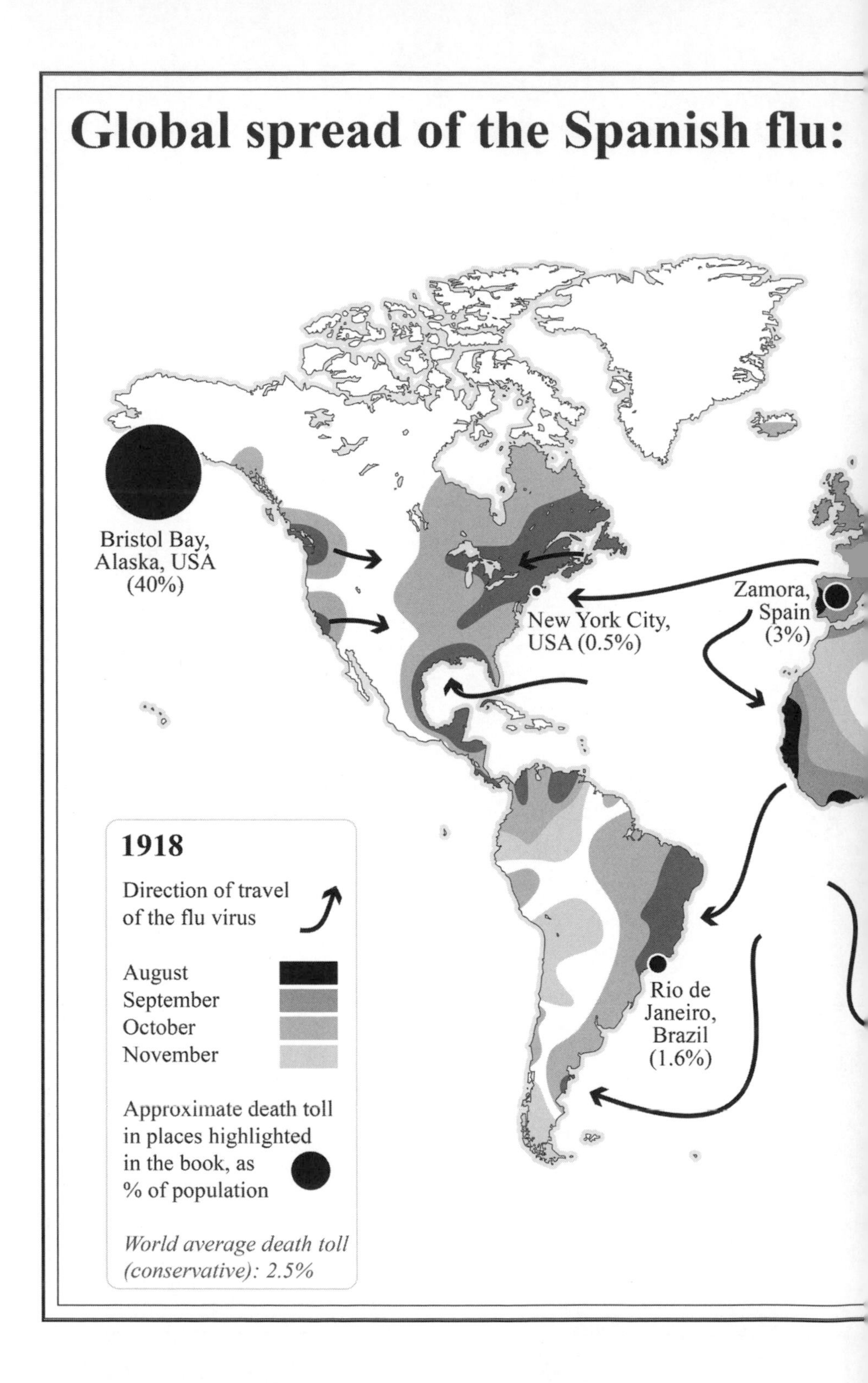
Global spread of the Spanish flu:
Bristol Bay, Alaska, USA (40%)
New York City, USA (0.5%)
Zamora, Spain (3%)
Rio de Janeiro, Brazil (1.6%)
1918
Direction of travel of the flu virus
August
September
October
November
Approximate death toll in places highlighted in the book, as % of population
World average death toll (conservative): 2.5%

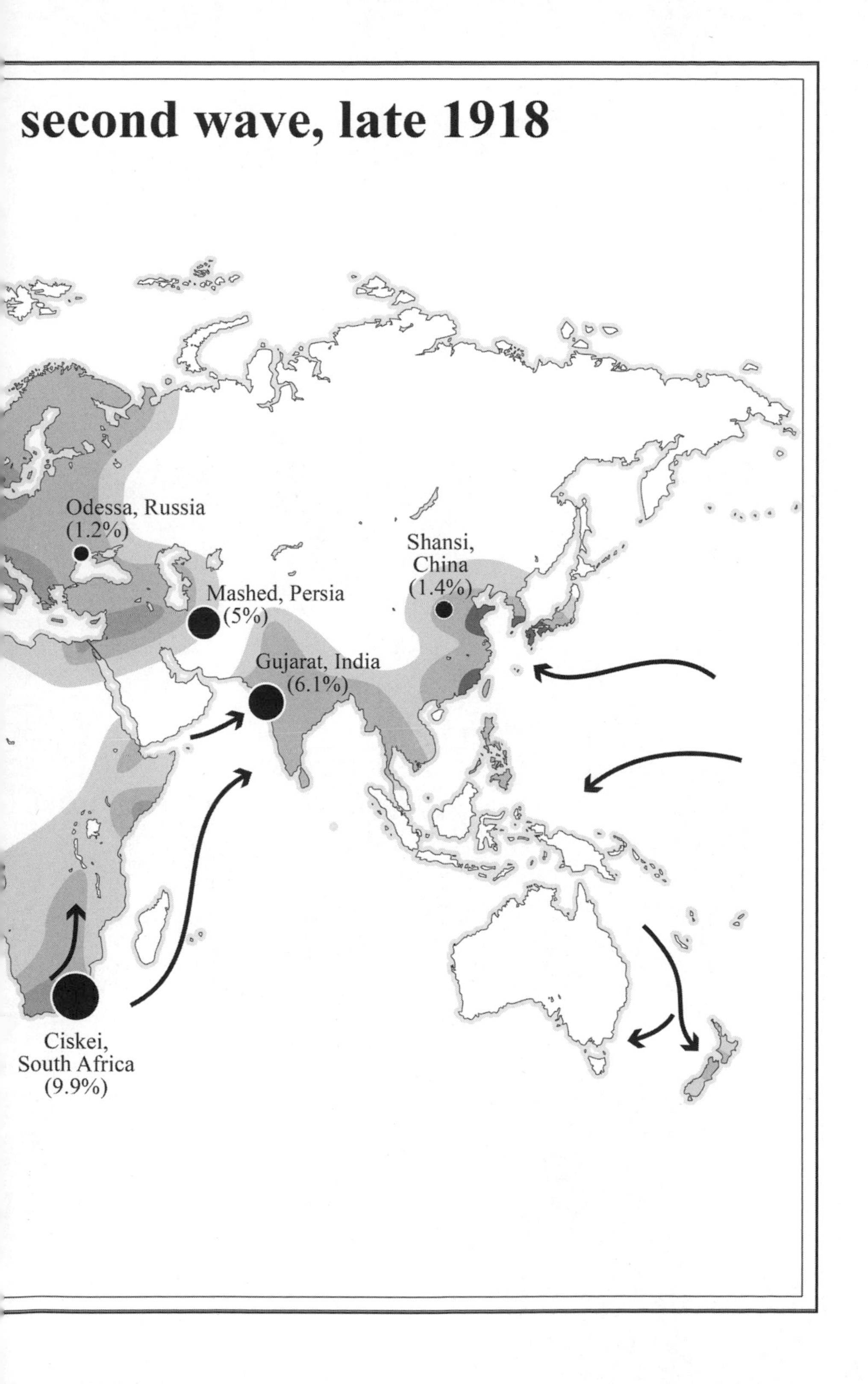
second wave, late 1918
Odessa, Russia
(1.2%)
Shansi,
China
(1.4%)
Mashed, Persia
(5%)
Gujarat, India
(6.1%)
Ciskei,
South Africa
(9.9%)

INTRODUCTION:

The Elephant in the Room

Japanese schoolgirls wearing protective masks during the pandemic, 1920

> The brevity of the influenza pandemic of 1918 posed great problems to doctors at the time . . . It has posed great problems to historians ever since.
>
> Terence Ranger, *The Spanish Influenza Pandemic of 1918–19* (2003)[1]

Kaiser Wilhelm of Germany abdicated on 9 November 1918 and in the streets of Paris there was jubilation. *'À mort Guillaume!'* they shouted. *'À bas Guillaume!'* Death to Wilhelm! Down with Wilhelm! Meanwhile, high above the city's seventh arrondissement, the poet Guillaume Apollinaire lay on his deathbed. A leading light of the French avant-garde movement, the man who invented the term 'surrealist' and inspired such figures as Pablo Picasso and Marcel Duchamp, he had signed up to fight in 1914. Having survived a shrapnel wound to the head and the drilling of a hole into his skull, he died of Spanish flu at the age of thirty-eight, and was declared *'mort pour la France'*.

His funeral was held four days later – two days after the armistice was signed. On leaving the Church of St Thomas Aquinas, the mourners set off eastwards towards Père Lachaise Cemetery. 'But as it reached the corner of Saint-Germain,' recalled Apollinaire's friend and fellow poet, Blaise Cendrars, 'the cortège was besieged by a crowd of noisy celebrants of the armistice, men and women with arms waving, singing, dancing, kissing, shouting deliriously the famous refrain of the end of the war: "No, you don't have to go, Guillaume. No you don't have to go . . .'" The famous refrain was directed with irony at the defeated kaiser, but it was filled with poignancy for the friends of Apollinaire.[2]

The poet's death serves as a metaphor for our collective forgetting of the greatest massacre of the twentieth century. The Spanish flu infected one in three people on earth, or 500 million human beings. Between the first case recorded on 4 March 1918, and the last sometime in March 1920, it killed 50–100 million people, or between 2.5 and 5 per cent of the global population – a range that reflects the uncertainty that still surrounds it. In terms of single events causing major loss of life, it surpassed the First World War (17 million dead), the Second World War (60 million dead) and possibly both put together. It was the greatest tidal wave of death since the Black Death, perhaps in the whole of human history.

Yet what do we see when we unravel the scroll of the twentieth century? Two world wars, the rise and fall of communism, perhaps some of the more spectacular episodes of decolonisation. We do not see the most dramatic event of them all, though it's right there before our eyes. When asked what was the biggest disaster of the twentieth century, almost nobody answers the Spanish flu. They're surprised by the numbers that swirl around it. Some become thoughtful and, after a pause, recall a great-uncle who died of it, orphaned cousins lost to sight, a branch of the family that was rubbed out in 1918. There are very few cemeteries in the world that, assuming they are older than a century, don't contain a cluster of graves from the autumn of 1918 – when the second and worst wave of the pandemic struck – and people's memories reflect that. But there is no cenotaph, no monument in London, Moscow or Washington DC. The Spanish flu is remembered personally, not collectively. Not as a historical disaster, but as millions of discrete, private tragedies.

Perhaps that has something to do with its shape. The First World War dragged on for four long years, but despite its name, the bulk of the action was concentrated in European and Middle

Eastern theatres. The rest of the world felt the hot wind sucking it into that vortex but remained outside it, and in some places the war seemed very far away indeed. The war had a geographical focus, in other words, and a narrative that unfolded in time. The Spanish flu, in contrast, engulfed the entire globe in the blink of an eye. Most of the death occurred in the thirteen weeks between mid-September and mid-December 1918. It was broad in space and shallow in time, compared to a narrow, deep war.

The African historian Terence Ranger pointed out in the early 2000s that such a condensed event requires a different storytelling approach. A linear narrative won't do; what's needed is something closer to the way that women in southern Africa discuss an important event in the life of their community. 'They describe it and then circle around it,' Ranger wrote, 'constantly returning to it, widening it out and bringing into it past memories and future anticipations.'[3] The Jewish text, the Talmud, is organised in a similar way. On each page, a column of ancient text is surrounded by commentaries, then by commentaries on the commentaries, in ever-increasing circles, until the central idea has been woven through space and time, into the fabric of communal memory. (There may be another reason why Ranger proposed a feminised history of the Spanish flu: it was generally women who nursed the ill. They were the ones who registered the sights and sounds of the sickroom, who laid out the dead and took in the orphans. They were the link between the personal and the collective.)

At the root of every pandemic is an encounter between a disease-causing microorganism and a human being. But that encounter, along with the events that lead up to it and the events that ensue from it, is shaped by numerous other events taking place at the same time – as well as by the weather, the price of bread, and ideas about germs, white men and jinns. The pandemic in its turn affects the price of bread, ideas about germs, white men and jinns – and sometimes even the weather. It is a social phenomenon as much as it is a biological one; it cannot be separated from its historical, geographical and cultural context.

The way African mothers and grandmothers recount an event gives weight to that contextual richness, even if the event it impinges on lasts no longer than a historical heartbeat. This book sets out to do the same.

The time is right. In the decades immediately after the pandemic, the only people who studied it – besides actuaries employed by insurance companies – were epidemiologists, virologists and medical historians. Since the late 1990s, however, Spanish-flu historiography has exploded, and this recent burst of attention has been noticeable by its multidisciplinary nature. Now economists, sociologists and psychologists are interested in it too, along with 'mainstream' historians. Each has trained their gaze on a different aspect, and between them they have transformed our understanding of it. All too often, though, their conclusions are buried in specialist journals, so this book attempts to bring them together; to weave all the different strands into a more coherent picture of the beast, in all its multifaceted glory – or horror.

The information that is available today is not only more academically diverse, it is also more geographically diverse – capturing the global reach of the disaster. Most accounts of the Spanish flu to date have focused on Europe or North America. They had to, because for a long time it was only in those places that data had been collected systematically. In 1998, when Spanish-flu experts from around the world met in Cape Town to mark its eightieth anniversary, they acknowledged that almost nothing was known about what happened in large swathes of the globe – South America, the Middle East, Russia, South East Asia and inland China. But Europe- and North America-centred accounts distort the picture for two reasons. First, those continents reported the lowest death rates, on average, so their experiences were atypical. And second, by 1918 they were both heavily implicated in a war that would devastate Europe. War was undoubtedly the main event on that continent: France lost six times more souls to the war than to the flu, while in Germany the multiple was four, in Britain three and in Italy two. But on *every other continent* – with the possible exception of Antarctica, which both disasters left pristine – more died of flu than war. At the time of writing – nearly twenty years on from the Cape Town summit, and as we approach the centenary of the catastrophe – it is possible to begin to reconstruct what happened in those other parts of the world.

This book takes a different approach to telling the flu. It moves in on it – from prehistory to 1918, from the planet to the human, from the virus to the idea and back again. At the heart of it is the story of how the Spanish flu emerged, swept the planet and receded, leaving humanity transformed. But that story pauses, at times, to look at what set communities apart in their experiences of it, as well as what brought them together. In 1918, the Italian-Americans of New York, the Yupik of Alaska and the residents of the Persian shrine city of Mashed had almost nothing in common except the virus, and in each place cultural and other factors moulded their encounter with it. A series of portraits therefore tracks the disaster as it unfolded in societies located at different points on the globe, highlighting the profoundly social nature of a pandemic.

These portraits shine a light into areas of the map that were previously dark, and give an indication of how the Spanish flu was experienced in parts of the world where 1918 was the year of the flu, not the year the war ended. They are not comprehensive, because millions of stories remain untold, so they come with caveats. It surely wasn't only in Rio de Janeiro that a post-flu orgy produced a spike in births, or only in Odessa, Russia, that people performed archaic religious rituals to ward off the scourge. It wasn't only Indians who temporarily transgressed strict social boundaries to help each other, or only in South Africa that people of one colour blamed those of another. A Catholic bishop may have frustrated efforts to contain the disease in Spain, but missionaries were often the only ones bringing relief to remote tracts of China. There is one overarching caveat: the narrator is, once again, European.

The story of the Spanish flu is told in Parts Two to Six of the book. But that story is embedded in a larger one – the one that tells how man and flu have cohabited, and co-evolved, for 12,000 years – so Part One, 'The Unwalled City', recounts that story up to 1918, while Part Seven, 'The Post-Flu World', explores the traces of the Spanish flu with which we live today. Since man and flu are still co-evolving, Part Eight, 'Roscoe's Legacy', looks forward to a future battle – the next flu pandemic – envisaging what new weapons we will carry into it, and what is likely to be our Achilles heel. Together, these stories comprise a biography of the flu – a human story, that is, in which the *fil conducteur* is flu. An afterword addresses the question of memory, asking why, when its impact was so profound, we call it 'forgotten'.

It is often said that the First World War killed Romanticism and faith in progress, but if science facilitated industrial-scale slaughter in the form of the war, it also failed to prevent it in the form of the Spanish flu. The flu resculpted human populations more radically than anything since the Black Death. It influenced the course of the First World War and, arguably, contributed to the Second. It pushed India closer to independence, South Africa closer to apartheid, and Switzerland to the brink of civil war. It ushered in universal healthcare and alternative medicine, our love of fresh air and our passion for sport, and it was probably responsible, at least in part, for the obsession of twentieth-century artists with all the myriad ways in which the human body can fail. 'Arguably' and 'probably' are indispensable qualifiers when discussing the Spanish flu, because in 1918 there was no way of diagnosing influenza, and hence no way of knowing for sure that that was what it was – any more than we can be certain that the bubonic plague (or one of its variants, pneumonic plague) caused the Black Death in the fourteenth century. What isn't arguable is that the 1918 pandemic accelerated the pace of change in the first half of the twentieth century, and helped shape our modern world.

If all of this is true, how come we still think of the Spanish flu as a footnote to the First World War? Have we really forgotten it? Terence Ranger thought we had, but if he were still alive today he might hesitate before repeating that claim. If so, then credit must go to a vast collaborative effort. The Spanish flu can no longer be told without the contributions of both historians and scientists, including social scientists. Science tells the tale up to the threshold of history, across the acres of prehistory that look empty but are, in fact, covered in an invisible scrawl – and that moulded events in 1918 just as much as what came later. History takes it up where the scrawl becomes legible, and science sheds some light back from the present. In another hundred years, science and history will themselves have been transformed. There might even be a science *of* history, in which theories about the past are tested against computerised banks of historical data.[4] That kind of approach will likely revolutionise the way we understand complex phenomena such as pandemics, but it's still in its infancy. There is one thing we can already say for sure, however: by the bicentenary of the 1918 pandemic, historians will have filled in more of the blanks, and the light shed by science will be brighter.

PART ONE: The Unwalled City

A crowded street scene in Bombay, circa 1920

I

Coughs and sneezes

Sometime around the winter solstice of 412 BC, a cough wracked the people of Perinthus, a port city on the Sea of Marmara in what was then northern Greece. The Perinthians reported other symptoms too: sore throat, aches, difficulty swallowing, paralysis of the legs, an inability to see at night. A doctor called Hippocrates jotted them all down, and the 'Cough of Perinthus' became the first written description – probably – of influenza.

Probably, because certain of those symptoms don't seem to fit: impaired night vision, paralysis of the limbs. Their inclusion troubled historians of medicine, until they realised that Hippocrates defined an epidemic differently from us. Indeed, Hippocrates was the first to use the word epidemic (literally, 'on the people') in a medical sense. Before that, it had referred to anything that propagates in a country, from fog to rumour to civil war. Hippocrates applied it specifically to disease, and then he redefined disease.

The ancient Greeks thought of disease as spiritual in origin, a punishment from the gods for any kind of misdemeanour. Doctors were part priests, part magicians, and it was their role to mollify the irascible divinities with prayer, spells and sacrifices. Hippocrates argued that the causes of disease were physical, and that they could be divined by observing a patient's symptoms. He and his disciples introduced a system for classifying diseases, which is why he is often referred to as the father of western medicine: he was responsible for the notions of diagnosis and treatment that still underpin medicine today (he also left us with a code of medical ethics, the Hippocratic Oath, from which we have the promise made by newly qualified doctors to 'do no harm').

Hippocrates thought that disease was the result of an imbalance between the four 'humours' or fluids that circulate in the human body – black bile, yellow bile, phlegm and blood. If you were lethargic, you had too much phlegm, and the treatment was to eat citrus fruit. Galen, another Greek physician who lived about 500 years after Hippocrates, elaborated on that model, suggesting that people could be categorised by temperament according to which humour dominated in them. Black bile was associated with melancholy types, yellow bile with choleric or hot-tempered ones. A phlegmatic person was laid-back, a sanguine one hopeful. We retain the adjectives, but not the understanding of anatomy and bodily function that produced them. And yet, the Galenic concept of medicine dominated in Europe for a good 1,500 years, and his notion that 'miasma' or noxious air could trigger a humoral imbalance was still popular, in some parts of the world, in the twentieth century.

Hippocrates' definition of an epidemic didn't survive either. For him, an epidemic was all those symptoms experienced in a given place over a given period of time, during which its population was in the grip of sickness. In those circumstances, he did not distinguish between separate diseases. Later the term epidemic came to be associated with one disease, then with one microbe, then with one strain of microbe, but this process of refinement didn't get underway until the Middle Ages, when the great plague epidemics forced a rethink. In modern terms, therefore, the people of Perinthus were probably suffering from influenza, diphtheria and whooping cough combined – perhaps with a deficiency of vitamin A thrown in.

Why should we care about a 2,400-year-old outbreak of flu in Greece? Because we would like to know how long flu has been a disease of humans, and what caused it to become one in the first place. Understanding more about its origins might help us to pinpoint the factors that determine the timing, size and severity of an outbreak. It might help us to explain what happened in 1918, and predict future epidemics.

The Cough of Perinthus probably wasn't the first flu epidemic. And though the historical record is silent on the subject before 412 BC, that doesn't mean there's nothing to be said about flu in earlier times. Like humans, flu carries information about its origins within itself. Both of us are living records of our evolutionary past. An example is the human tail bone or coccyx, which is a vestige of our tree-dwelling ancestors. As the tail became less useful, natural selection favoured individuals in whom a chemical signal during embryonic development switched off spinal elongation before the tail grew. Very occasionally, a glitch occurs and that signal doesn't get turned off in time. The medical literature contains around fifty reports of babies born with tails – a glimpse of the arboreal primate in all of us.

The flu virus has no tail, but it harbours other clues to its origins. It is a parasite, meaning that it can only survive inside another living organism, or 'host'. Unable to reproduce on its own, it must invade a host cell and hijack that cell's reproductive apparatus. The offspring of the virus must then leave that host and infect a new one. If they don't, then the virus expires with the original host, and that is the end of flu. Just as our ancestors' survival depended on their ability to swing through trees, so flu's survival depends on its ability to jump from one host to another. This is where the flu story becomes interesting, however, because being a parasite, its survival depends both on its own behaviour *and on that of its host*. And though for a long time scientists were in the dark about flu's past, they knew a few things about what humans were doing before 412 BC.

Flu is transmitted from one person to another in tiny infected droplets of mucus that are flung through the air by coughs and sneezes. Snot is a fairly effective missile – it should be, it was designed in a wind tunnel – but it can't fly further than a few metres. For flu to spread, therefore, people must live fairly close together. This was a crucial insight, because people didn't always live close together. For most of the human story they were hunter-gatherers and far apart. That all changed about 12,000 years ago, when a hunter somewhere in the vastness of Eurasia erected a pen around a couple of wild sheep and invented livestock. Plants were domesticated too, for crops, and these two developments meant that the land could now support a higher density of people, who could thus come together to compete, collaborate, and generally display all the ingenuity characteristic of human societies. The hunter's innovation, known as the farming revolution, ushered in a new era.

The new collectives that farming supported gave rise to new diseases – the so-called 'crowd diseases' such as measles, smallpox, tuberculosis and influenza. Humans had always been susceptible to infectious disease – leprosy and malaria were causing misery long before the farming revolution – but these were adapted to surviving in small, dispersed human populations. Among their tricks for doing so were not conferring total immunity on a recovered host, so that he or she could be infected again, and retreating to another host – a so-called 'animal reservoir' – when humans were scarce. Both strategies helped ensure that they maintained a sufficiently large pool of susceptible hosts.

The crowd diseases were different. They burned rapidly through a farming population, either killing their victims or leaving them immune to re-infection. They might infect other animals, but not as well as they infected humans, and some of them were so well adapted to humans that they became exclusively parasitic to our species. They needed a pool of thousands or even tens of thousands of potential victims to sustain them – hence the name, 'crowd disease'. They would not have survived prior to the farming revolution, but after it, their evolutionary success was index-linked to the growth of human populations.

But if they would not have survived before farming, where did they come from? The clue is those animal reservoirs. We know that there are disease-causing microbes that only infect animals. There are forms of malaria, for example, that infect birds and reptiles but can't be transmitted to humans. We know that there are microbes that infect both animals and humans (influenza falls into this category), and we know that there are microbes that infect only humans. This is the case, for example, with measles, mumps and rubella. According to current thinking, these different categories of infectious disease represent steps on the evolutionary path by which an exclusively animal disease becomes an exclusively human one. To be precise, scientists recognise five steps that a disease-causing microbe has to go through to complete this transition.[1] Some diseases, like measles, have gone all the way; others have stuck at intermediate points on the path. But we shouldn't think of this process as fixed. It's highly dynamic, as illustrated by Ebola.

Ebola virus disease is primarily a disease of animals. Its natural reservoir is thought to be fruit bats that inhabit African forests, and that may infect other forest-dwelling animals that humans prize as bushmeat (humans eat the bats, too). Until recently, Ebola was considered a disease that infected humans poorly: it might be transmitted via contact with bushmeat, for example, but a person who was infected by that route would only infect a few others before the 'outbreak' fizzled out. That all changed in 2014, when an epidemic in West Africa revealed that Ebola had acquired the ability to pass easily between people.

It isn't easy for, say, a virus to jump the species barrier. In fact 'jump' is entirely the wrong word – it would be more helpful, though still a metaphor, to think of it 'oozing' across. Cells are built differently in different hosts, and invading them requires different tools. Each step along the path to becoming a human disease is therefore accompanied by a specific set of molecular changes, but acquiring those changes is a very hit-and-miss affair. The virus will likely have to pass through many, many rounds of reproduction before a mutation arises that confers a useful change. But then, if the virus's evolutionary fitness improves as a result – if by infecting humans better, it manages to produce more of itself – then natural selection will favour that change (if it doesn't, it won't). Other changes may follow, and their cumulative effect is that the virus moves another step along the path.

The natural reservoir of influenza is generally considered to be birds, especially waterbirds. The big giveaway that a certain species plays the role of reservoir for a certain pathogen is that it doesn't get sick from it. The two have co-evolved for so long that the virus manages to complete its life cycle without causing too much damage to its host, and without unleashing an immune response. Ducks, for example, can be heavily infected with flu without showing any signs of disease. After the farming revolution, ducks were among the animals that humans domesticated and brought into their villages. So were pigs, which are regarded as potential intermediaries in the process by which a bird disease became a human disease, since pig cells share features of both human and bird cells. For millennia, the three lived cheek by jowl, providing flu with the ideal laboratory in which to experiment with moving between species. Flu infected humans, but probably not very well at first. Over time, however, it accumulated the molecular tools it needed to make it highly contagious, and one day there was an outbreak deserving of the name 'epidemic'.

Epidemic here is meant in its modern sense – that is, as an increase, often sudden, in the number of cases of a given disease in a given population. An 'endemic' disease, in contrast, is always found in that population. A crowd disease can be both endemic and epidemic, if it is always present in a region but also produces occasional outbreaks there. This is where the definitions of the two terms become a little blurred and vary according to the disease in question. We might say, for example, that the relatively mild outbreaks of seasonal influenza that we see each winter are the endemic form of the disease, and reserve the term epidemic for when a new strain emerges, bringing a more severe form of flu in its wake – though not everybody would agree with that distinction.

We have no written accounts of the first epidemics of the first crowd diseases, but they are likely to have been very deadly (witness the 2014 epidemic of Ebola, which might yet go on to earn the title 'crowd disease'). We know, for example, that one of the deadliest crowd diseases of all, smallpox, was present in Egypt at least 3,000 years ago, because mummies have been found with pockmarked faces, but the first written account of a (probable) smallpox epidemic doesn't turn up until 430 BC, when a contemporary of Hippocrates, Thucydides, described corpses piled up in the temples of Athens.

When did the first flu epidemic occur? Almost certainly in the last 12,000 years, and probably in the last 5,000 – since the first cities arose, creating ideal conditions for the disease to spread. It too must have been horrific. We find this hard to understand, because today, in general, influenza is far from lethal. Yet even today, a small proportion of people come off badly each flu season. These unlucky individuals develop acute respiratory distress syndrome (ARDS): they become short of breath, their blood pressure drops, their faces take on a bluish tinge, and if they aren't rushed to hospital they will very likely die. In a few cases, their lungs may even haemorrhage, causing them to bleed from their noses and mouths. ARDS is a glimpse of the carnage that first flu epidemic wrought.

There is no record of it (the oldest full writing system wasn't developed until 4,500 years ago), so we don't know when or where it happened, but Uruk in what is now Iraq might be a good candidate. Considered the largest city in the world 5,000 years ago, Uruk had around 80,000 inhabitants living inside a walled enclosure of six square kilometres – twice the area of London's financial heart, the City. Nobody had any immunity. Nobody could help anybody else. Many would have died. Other flu epidemics must have followed, and they were probably milder: though the strains that caused them differed from that original one, and from each other, they were similar enough that the survivors gradually acquired some immunity. Influenza gradually came to look more like the disease we recognise today, though at the cost of a great many lives.

'Against other things it is possible to obtain security,' wrote the Greek philosopher Epicurus in the third century BC, 'but when it comes to death we human beings all live in an unwalled city.'[2] From the moment influenza became a human disease, it began to shape human history – though we had to wait for Hippocrates to write the first (probable) description of it. Even after Hippocrates, it's hard to be sure that what is being described is influenza as we know it. Not only have concepts of epidemic and disease changed, but the disease itself has gone by different names, reflecting changing ideas about what causes it. On top of that, flu is easily confused with other respiratory diseases – most obviously the common cold, but also more serious diseases such as typhus and dengue fever, that start out with flu-like symptoms.

Treading carefully, aware of the traps that time inserts between words, historians have nevertheless speculated that it was flu that devastated the armies of Rome and Syracuse in Sicily in 212 BC. 'Deaths and funerals were a daily spectacle,' wrote Livy in his *History of Rome*. 'On all sides, day and night, were heard the wailings for the dead.'[3] It may have been the respiratory disease that raged through Charlemagne's troops in the ninth century AD, that he knew as *febris Italica* (Italian fever). Probable flu epidemics were documented in Europe in the twelfth century, but the first really reliable description of one doesn't appear until the sixteenth century. In 1557, in the brief interlude when Mary I was on the English throne, an epidemic eliminated 6 per cent of her subjects – more Protestants than 'Bloody Mary', as she became known, could dream of burning at the stake.

By the sixteenth century, the age of discovery was well underway. Europeans were arriving in ships in the New World, bringing with them their newfangled diseases to which local populations had no immunity. They had no immunity because they had not been through the same harrowing but tempering cycle of epidemics of animal origin. The fauna of the New World lent itself less easily to domestication than that of the Old, and some inhabitants were still hunter-gatherers. Flu may have been the disease that travelled with Christopher Columbus on his second voyage to the New World, in 1493, and that wiped out much of the Amerindian population of the Antilles after he stopped off there. That year, the Caribbean experienced something similar to what happened, several millennia earlier, in a Eurasian city like Uruk – only this time one group was left standing: the *conquistadors*.

For a long time historians ignored infectious diseases as historical players, not suspecting this imbalance in their effects on different populations. Right up until the twentieth century, European historians recounting Spaniard Hernán Cortés' astonishing David-and-Goliath conquest of the Aztec Empire in Mexico generally failed to mention that an epidemic of smallpox did most of the work for him.[4] For them, flu was a mild irritant, a cross to be borne in the darker months. They didn't grasp the fear it struck into the hearts of Native Americans, Australians or Pacific Islanders, or how closely those peoples associated it with the coming of the white man. 'There was a firm belief among all, that of late years, since they had visits from white men, their influenza epidemics were far more frequent and fatal than they used to be,' wrote one nineteenth-century visitor to Tanna in the Vanuatu archipelago. 'This impression is not confined to Tanna, it is, if I mistake not, universal throughout the Pacific.' Once the historians had realised their error, some of them started calling crowd diseases by a different name: imperial diseases.[5]

It was the work of palaeoclimatologists that brought that error home to them. Palaeoclimatologists try to understand what the earth's climate was like in the past, and why, by studying such things as sediment deposits, fossils and tree rings. Finding that the world grew cooler in the late Roman era, for example, they suggest that the Plague of Justinian – a pandemic of bubonic plague that killed approximately 25 million people in Europe and Asia in the sixth century AD – led to vast tracts of farmland being abandoned and forests growing back. Trees extract carbon dioxide from the atmosphere, and this reforestation led to so much of the gas being sequestered in wood that the earth cooled (the opposite of the greenhouse effect we are witnessing today).

Similarly, the massive waves of death that Cortés, Francisco Pizarro (who conquered the Inca Empire in Peru) and Hernando de Soto (who led the first European expedition into what is now the United States) unleashed in the Americas in the sixteenth century caused a population crash that may have ushered in the Little Ice Age.[6] The effect wasn't reversed until the nineteenth century, when more Europeans arrived and began to clear the land again. The Little Ice Age was probably the last time a human disease affected the global climate, however. Though there would be other pandemics, the gradual mechanisation of farming, combined with the exponential growth of the world's population, meant that even the deaths of tens of millions of farmers could leave no dent in the atmosphere – at least not one that palaeoclimatologists have been able to detect.

The first flu pandemic that experts agree was a pandemic – that is, an epidemic that encompassed several countries or continents – is thought to have begun in Asia in 1580, and spread to Africa, Europe and possibly America. Here, though, we have to introduce a caveat. Determining the origin and direction of spread of a flu pandemic is not easy – as we'll see – meaning that every categorical statement regarding the source of historic flu pandemics should be taken with a pinch of salt. This is especially true since, from at least the nineteenth century, Europeans whose compatriots had once tracked lethal diseases through the New World were quick to see each new plague as blowing out of China, or the silent spaces of the Eurasian steppes.

Contemporary reports suggest that this first flu pandemic spread from north to south across Europe in six months. Rome recorded 8,000 deaths, meaning that it was literally 'decimated' – roughly one in ten Romans died – and some Spanish cities suffered a similar fate.[7] Between 1700 and 1800 there were two flu pandemics. At the height of the second, in 1781, 30,000 people a day were falling sick in St Petersburg. By then, most people were calling the disease 'influenza'. The name was first coined by some fourteenth-century Italians who attributed it to the pull or 'influence' of the stars, but it took several centuries to catch on. We retain it today, of course, though as with the descriptors 'melancholy' and 'phlegmatic', its conceptual moorings have been swept away.

It was in the nineteenth century that crowd diseases reached the zenith of their evolutionary success, and held dominion over the globe. This was the century of the Industrial Revolution, and accompanying it, the rapid expansion of cities in many parts of the world. These cities now became breeding grounds for crowd diseases, such that urban populations were unable to sustain themselves – they needed a constant influx of healthy peasants from the countryside to make up for the lives lost to infection. Wars, too, brought epidemics in their wake. Conflict makes people hungry and anxious; it uproots them, packs them into insanitary camps and requisitions their doctors. It makes them vulnerable to infection, and then it sets large numbers of them in motion so that they can carry that infection to new places. In every conflict of the eighteenth and nineteenth centuries, more lives were lost to disease than to battlefield injuries.

The nineteenth century saw two flu pandemics. The first, which erupted in 1830, is said to have ranked in severity – though not in scale – with the Spanish flu. The second, the so-called 'Russian' flu that began in 1889, was thought to have originated in Bokhara in Uzbekistan. It was the first to be measured, at least to some extent, since by then scientists had discovered what a powerful weapon statistics could be in the fight against disease. Thanks to the efforts of those early epidemiologists, we know that the Russian flu claimed somewhere in the region of a million lives, and that it washed over the world in three waves. A mild first wave heralded a severe second one, and the third was even milder than the first. Many cases developed into pneumonia, which was often the cause of death, and this flu didn't only claim the elderly and the very young – as in a normal flu season – but people in their thirties and forties too. Doctors were unsettled by their observation that many patients who survived the initial attack went on to develop nervous complications, including depression. The Norwegian artist Edvard Munch may have been one of them, and some have suggested that his famous painting, *The Scream*, sprang from his flu-darkened thoughts. 'One evening I was walking along a path, the city was on one side and the fjord below,' he wrote later. 'I felt tired and ill. I stopped and looked out over the fjord – the sun was setting, and the clouds turning blood red. I sensed a scream passing through nature; it seemed to me that I heard the scream.'[8] By the time Munch wrote those words, the pandemic was over, and so was the millennia-long struggle between man and flu. In the next century, the twentieth, science would conquer the crowd diseases once and for all.

2

The monads of Leibniz

To us living in a world a hundred years older, a world in the grip of an AIDS pandemic, the idea that science would conquer infectious diseases for good seems nonsensical. But at the turn of the twentieth century many people believed it, at least in the west. The main reason for their optimism was germ theory – the insight that germs cause disease. Bacteria had been known about for a couple of centuries, ever since a Dutch lens grinder named Antony van Leeuwenhoek passed a magnifying glass over a drop of pond water and saw that it was teeming with life, but they had been regarded as a kind of harmless ectoplasm – nobody suspected that they could make people ill. Robert Koch in Germany and Louis Pasteur in France made the connection, starting in the 1850s. The discoveries of these two men are too numerous to list, but among them, Koch showed that TB, the 'Romantic' disease of poets and artists, was not inherited – as was widely believed – but caused by a bacterium, while Pasteur disproved the notion that living organisms could be generated spontaneously from inanimate matter.

In combination with older ideas about hygiene and sanitation, germ theory now began to turn the tide on the crowd diseases. Campaigns were launched to purify drinking water and promote cleanliness. Vaccination programmes were imposed, though not without resistance – not surprisingly, people balked at the idea that they could be protected against a disease by being injected with it – and these efforts produced concrete results. If in the wars of previous centuries, more lives had been lost to disease than to combat, that trend was now reversed. Weapons had become more lethal, but military doctors had also become better at controlling infection. Those might seem like odd grounds for claiming success, but army doctors were among the first to put germ theory into practice, and their expertise trickled down to their civilian counterparts. At the beginning of the twentieth century, cities at last became self-sustaining.

In the early decades of that century, therefore, faith in science and rationalism was high. The excitement over the discovery of the link between bacteria and disease had not yet abated, and there was a temptation to find bacteria responsible for every malaise. Ilya Mechnikov, the wild Russian 'demon of science' whom Pasteur had brought to his institute in Paris, even blamed them for old age. Mechnikov had won a Nobel Prize in 1908 for his discovery of phagocytosis – the mechanism by which immune cells in human blood swallow up harmful bacteria and destroy them. But he also suspected bacteria in the human intestine of releasing toxins that harden the arteries, contributing to the body's ageing – a belief that brought a certain amount of ridicule down on his head. He became obsessed with villages in Bulgaria where people reputedly lived to be more than a hundred, attributing their longevity to the sour milk they drank – and in particular, to the 'good' bacteria that soured it. In the last years of his life, he drank huge quantities of sour milk, before dying in 1916 at the age of seventy-one.[1] (These days, the microbes in our gut are generally considered to be either harmless, or good for us.)

Viruses, however, were still a mystery. In Latin the word *virus* means something like poison, or potent sap, and at the turn of the twentieth century that was exactly how people understood it. When in his 1890 novel *O Cortiço* (*The Slum*), Brazilian writer Aluísio Azevedo wrote 'Brazil, that inferno where every budding flower and every buzzing bluebottle fly bears a lascivious virus', a venomous secretion is probably what he had in mind. But scientists were beginning to question that definition. Were they toxins or organisms? Liquid or particle? Dead or alive? The first virus was discovered in 1892, when Russian botanist Dmitri Ivanovsky identified a virus as the cause of a disease in tobacco plants. He hadn't seen it. What he had discovered was that the disease was caused by an infectious agent that was smaller than all known bacteria – too small to see.

In 1892, the Russian flu was raging across Europe, and it was in the same year that Ivanovsky made his discovery that a student of Koch, Richard Pfeiffer, identified the bacterium responsible for influenza. That's right, the *bacterium* responsible for influenza. Pfeiffer's bacillus, also known as *Haemophilus influenzae*, really exists, and it causes disease, but it does not cause flu (Pfeiffer's error lives on in its name, like a warning to scientists, or a bad historical joke). Nobody suspected that flu could be the work of a virus, that unclassifiable thing that existed somewhere beyond the limits of observability, and they continued to not suspect it in 1918. In fact, viruses occupied only a tiny corner of the psychic universe of 1918. They hadn't been seen, and there was no test for them. These two facts are crucial to understanding the impact of the Spanish flu. Things changed in the wake of the pandemic, as this book will explain, but it took time. When James Joyce wrote, in his thoroughly modern novel *Ulysses* (1922), 'Foot and mouth disease. Known as Koch's preparation. Serum and virus', he probably thought of a virus in much the same way as Azevedo had.[2]

The disciples of Pasteur and Koch disseminated germ theory far and wide, so that it gradually displaced Galenic concepts of disease. The psychological shift that this demanded was as troubling as the one Hippocrates had provoked more than 2,000 years earlier, and people were slow to embrace it. When two waves of cholera swept London in the mid-nineteenth century, its residents blamed miasma rising from the filthy River Thames. After a brilliant piece of detective work that involved marking fatal cases of the disease on a map, a doctor called John Snow traced the source of one outbreak to a particular water pump in the city, and deduced – correctly – that water rather than air spread cholera. He published his conclusion in 1854, but it was only after the 'Great Stink' of 1858 – when a spell of hot weather rendered the smell of untreated sewage on the banks of the Thames overpowering – that the authorities finally commissioned an engineer, Joseph Bazalgette, to design a proper system of sewers for the city. Their reasoning? By eliminating the miasma, they would eliminate cholera too.

Germ theory also had profound implications for notions of personal responsibility when it came to disease. Hippocrates had some surprisingly modern ideas about this. People were responsible for their diseases, he believed, if they did not make lifestyle choices conducive to good health, but they could not be blamed if a disease was hereditary. Even in that case, however, they had choices. He gave the example of cheese, arguing that one should choose whether or not to eat cheese in the light of knowledge about the constitution one had inherited. 'Cheese,' he wrote, 'does not harm all men alike; some can eat their fill of it without the slightest hurt, nay, those it agrees with are wonderfully strengthened thereby. Others come off badly.'[3]

By the Middle Ages, people had shifted most of the responsibility for disease back onto the gods, or God, and a sense of fatalism persisted for centuries, despite the rise of science. In 1838, the French writer George Sand took her tubercular lover Frédéric Chopin to the Spanish island of Majorca, hoping that the Mediterranean climate would ease the symptoms of her 'poor melancholy angel'. She didn't expect it to cure him, because to her mind TB was incurable. Nor did it occur to her that she could catch it from him. By then, however, ideas about what caused TB were already in flux, and when the pair arrived in Palma, they discovered that its inhabitants wanted nothing to do with them. As an outraged Sand wrote to a friend, they were asked to leave, TB 'being extremely rare in those latitudes and, moreover, considered contagious!'[4]

In the nineteenth century, epidemics were still regarded – like earthquakes – as acts of God. Germ theory forced people to consider the possibility that they could control them, and this revelation brought another new set of ideas into play: the theory of evolution that Charles Darwin had introduced in his *On the*

Origin of Species (1859). When Darwin had talked about natural selection, he had not meant his ideas to be applied to human societies, but his contemporaries did just that, giving birth to the 'science' of eugenics. Eugenicists believed that humanity comprised different 'races' that competed for survival. The fittest thrived, by definition, while the 'degenerate' races ended up living in poverty and squalor because they lacked drive and self-discipline. This line of thinking now dovetailed insidiously with germ theory: if the poor and the working classes also suffered disproportionately from typhus, cholera and other killer diseases, then that too was their fault, since Pasteur had taught that such diseases were preventable.

Eugenics informed immigration and public health policies across the world in the late nineteenth century. German anthropologists were busy classifying human 'types' in their colonies in Africa, while in some American states, people judged mentally ill were forcibly sterilised. Ironically, though American eugenicists saw the Japanese as racially inferior, and tried to keep them out of their country, eugenics was also popular in Japan – where, of course, the Japanese race was regarded as superior.[5] Eugenics is taboo today, but in 1918 it was mainstream, and it would powerfully shape responses to the Spanish flu.

'The minds of different generations are as impenetrable one by the other as are the monads* of Leibniz,' wrote Frenchman André Maurois, but we can at least highlight some obvious differences between 1918 and now. The world was at war, and had been since 1914. The reasons for that war lay mainly in Europe – in tensions between that continent's great imperial powers. The age of discovery had borne fruit by 1914, when more of the globe was colonised by Europeans than at any other time. From that apogee, a long process of decolonisation would break up those empires and liberate their colonies. But 1918 also saw one of the

* A monad, according to the German philosopher Gottfried Leibniz (1646–1716), was an elementary particle or simple substance – something indivisible.

last battles in one of the last colonial wars – the American Indian Wars, in which the European settlers of North America fought, and ultimately defeated, its indigenous peoples.

Future heads of state Nicolae Ceauşescu and Nelson Mandela were born in 1918, as was the future dissident writer Aleksandr Solzhenitsyn, the film director Ingmar Bergman and the actress Rita Hayworth. Max Planck won the Nobel Prize in Physics for his work on quantum theory, while Fritz Haber won the chemistry prize for inventing a way of producing ammonia, which is important in the manufacture of fertilisers and explosives (the Nobel committees decided not to award prizes in medicine, literature or peace that year). Gustav Holst's *The Planets* was lauded at its premiere in London, while Joan Miró's work was lampooned at the artist's first solo exhibition in Barcelona.

Movies were silent and telephones were rare. Long-distance communication was mainly by telegraph, or, in parts of China, by carrier pigeon. There were no commercial airplanes, but there were submarines, and steamships plied the oceans at an average speed of a little under twelve knots (about twenty kilometres per hour).[6] Many countries had well-developed rail networks, but many did not. Persia, a country three times the size of France, had twelve kilometres of rail. It also had only 300 kilometres of road and a single car – the shah's. Ford had issued his affordable Model T, but cars were a luxury, even in America. The most common mode of transport was the mule.

It was a world that was both familiar to us, and terribly foreign. Despite the inroads made by germ theory, for example, human populations were far less healthy than they are now, and even in the industrialised world, the main cause of ill health was still, overwhelmingly, infectious diseases – not the chronic, degenerative diseases that kill most of us today. After America entered the war in 1917, it undertook a mass examination of army draftees – the first national physical exam in its history. The results came to be known as 'the horrible example': of 3.7 million men who were examined, around 550,000 were rejected as unfit, while almost half the remainder were found to have some physical deformity – often one that was preventable or curable.

'Plague', for us, means something very precise: bubonic plague – as well as its variants, pneumonic and septicaemic plague – all of which are caused by the bacterium *Yersinia pestis*. But in 1918, 'plague' referred to any dangerous disease that attacked by storm. 'Real' plague, meanwhile – the disease that, under the alias 'the Black Death', devastated medieval Europe – was still present on that continent. It seems extraordinary, but in England its last visitation coincided with that of the Spanish flu.[7] Nor did 'middle-aged' mean what it does today: life expectancy at birth in Europe and America did not exceed fifty, and in large parts of the globe it was much lower. Indians and Persians, for example, were lucky to celebrate their thirtieth birthdays.

Even in wealthy countries, the vast majority of births took place at home, bathtubs were reserved for the rich, and a significant minority of the population was illiterate. Ordinary people grasped the concept of contagion, but not the mechanism, and if that seems surprising – given that germ theory had already been around for half a century by then – consider a modern parallel. The discovery of the structure of DNA in 1953 gave birth to the field of molecular genetics, which once again radically altered our understanding of health and disease. But a survey of ordinary Americans conducted half a century later, in 2004, revealed high levels of confusion as to what a gene actually is.[8]

Doctors' training was patchy in 1918, though from 1910 Abraham Flexner had begun campaigning for rigorous, standardised medical education in the US. Health insurance was almost unheard of, and in general healthcare was paid for privately or provided by charities. Antibiotics had yet to be invented, and there was still relatively little people could do once sick. Even in Paris and Berlin, therefore, disease filled the interstices of human lives. It lurked behind the column inches devoted to the war. It was the dark matter of the universe, so intimate and familiar as not to be spoken about. It engendered panic, followed by resignation. Religion was the main source of comfort, and parents were used to surviving at least some of their children. People regarded death very differently. It was a regular visitor; they were less afraid.

This, then, was the world into which the Spanish flu erupted: a world that knew the motor car but was more comfortable with the mule; that believed in both quantum theory and witches; that straddled the modern and premodern eras, so that some people lived in skyscrapers and used telephones, while others lived much as their ancestors had in the Middle Ages. There was nothing modern about the plague that was about to be unleashed on them, however; it was thoroughly ancient. From the first fatality, it was as if the entire population of the globe, some 1.8 billion people, had been transported back several millennia, to a city like Uruk.

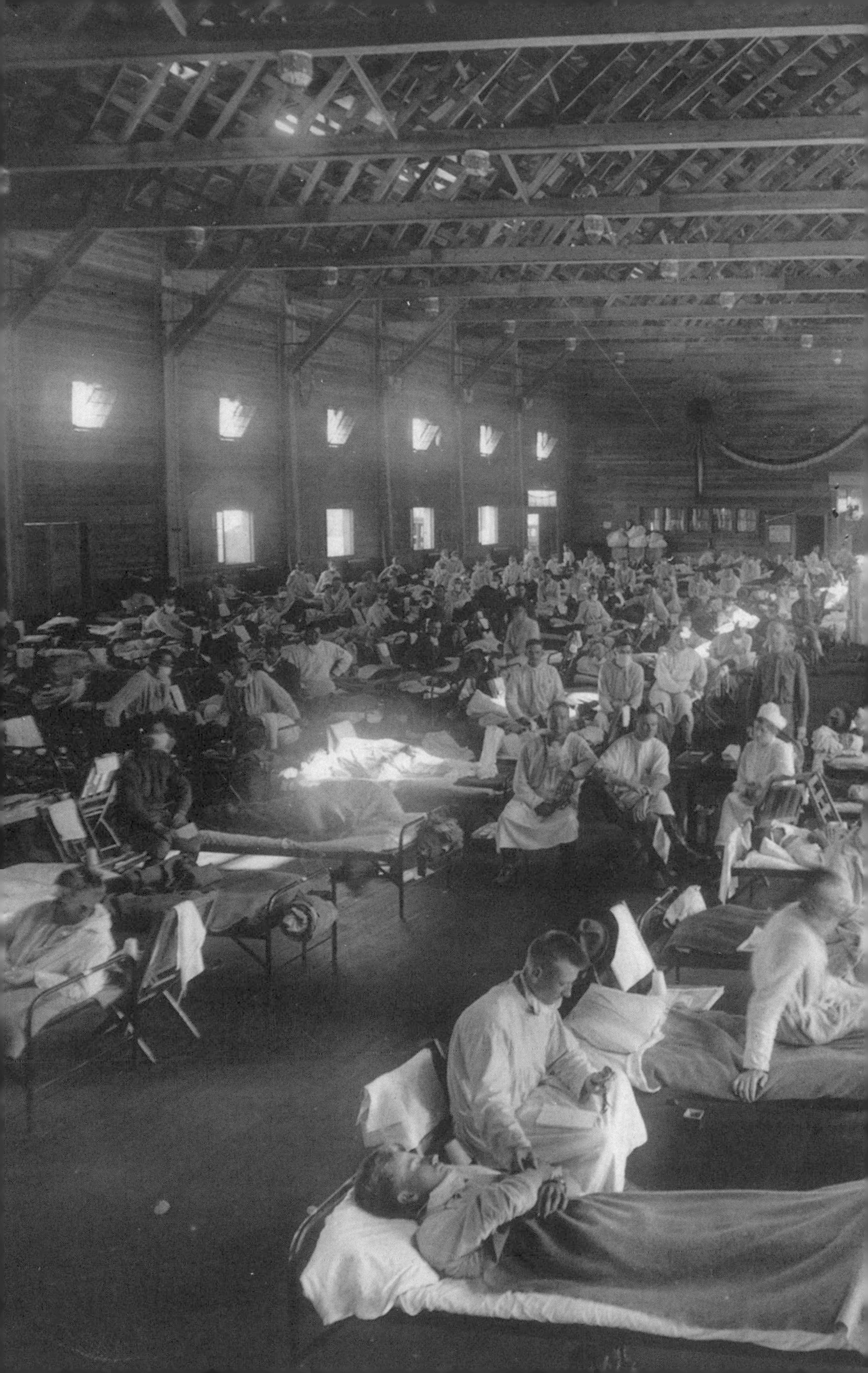

PART TWO: Anatomy of a Pandemic

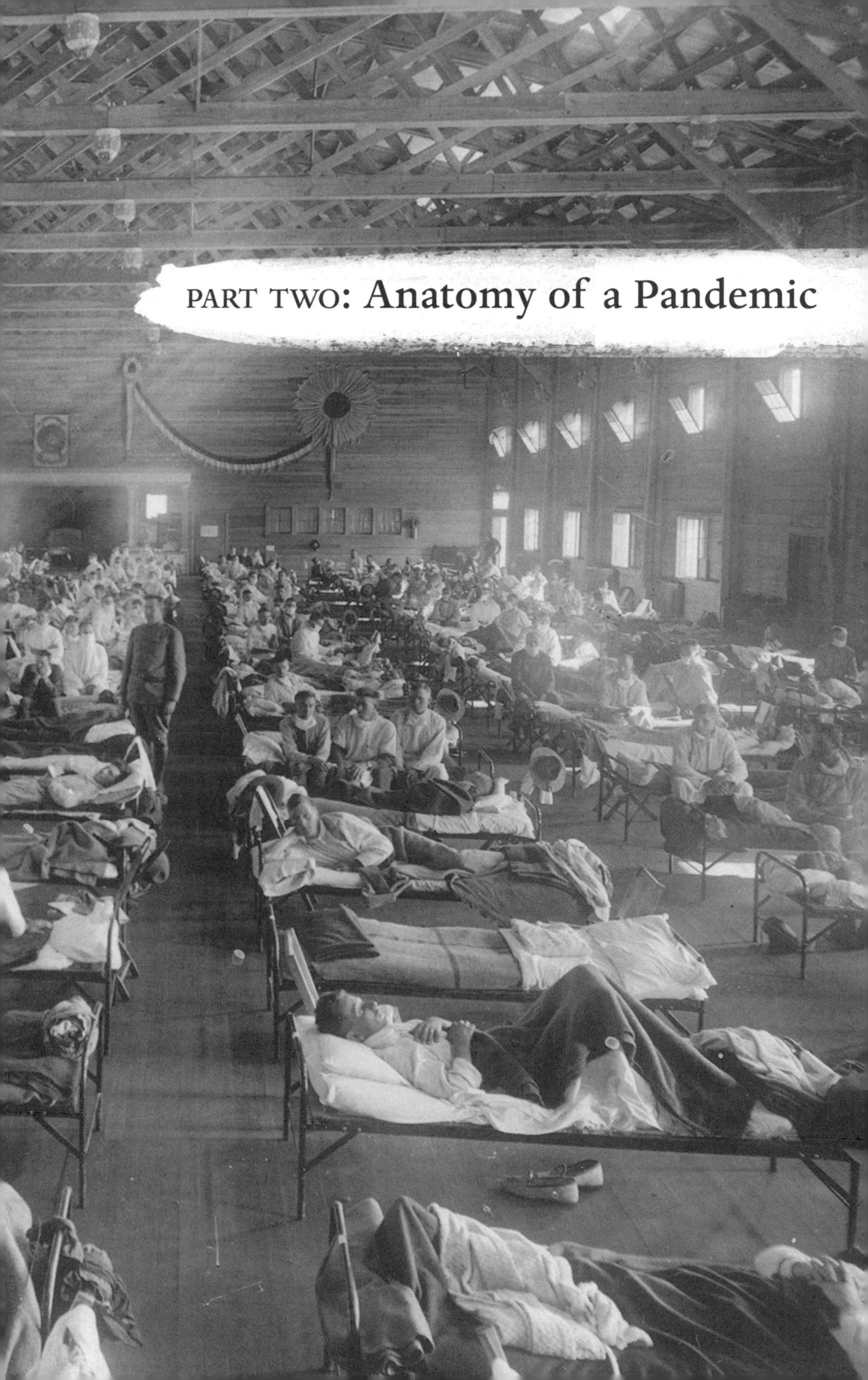

Emergency hospital created to accommodate Spanish flu patients at the US Army's Camp Funston, Kansas, 1918

3

Ripples on a pond

On the morning of 4 March 1918, a mess cook at Camp Funston in Kansas, Albert Gitchell, reported to the infirmary with a sore throat, fever and a headache. By lunchtime the infirmary was dealing with more than a hundred similar cases, and in the weeks that followed so many reported sick that the camp's chief medical officer requisitioned a hangar to accommodate them all.

Gitchell may not have been the first person to catch the 'Spanish' flu. Beginning in 1918, and continuing right up until the present day, people speculated about where the pandemic had actually started. We now know, however, that his case was among the first to be officially recorded, and so by consensus – for the sake of convenience – it is generally considered to mark the beginning of the pandemic. Five hundred million others would follow Albert, metaphorically speaking, to the infirmary.

America had entered the First World War in April 1917, and that autumn, young men from mainly rural parts of the country began to converge on military camps to be recruited and trained for the American Expeditionary Forces (AEF) – the armed forces that General John 'Black Jack' Pershing would lead into Europe. Camp Funston was one such camp. It supplied soldiers both to other American camps and directly to France. By April 1918, the flu was epidemic in the American Midwest, on the cities of the eastern seaboard from which the soldiers embarked, and in the French ports where they disembarked. By the middle of April it had reached the trenches of the Western Front. The weather in western Europe was unseasonably hot that month, but German troops were soon complaining of *Blitzkatarrh*, something that no doubt preoccupied the German Second Army's director of hygiene, Richard Pfeiffer – the man who had lent his name to Pfeiffer's bacillus. From the front it quickly spread to the whole of France, and from there to Britain, Italy and Spain. Towards the end of May, the Spanish king, Alfonso XIII, fell sick in Madrid, along with his prime minister and members of his Cabinet.[1]

Also in May, flu was reported in Breslau in Germany, now Wrocław in Poland (where Pfeiffer held the chair of hygiene in peacetime), and in the Russian port of Odessa, 1,300 kilometres to the east. After Russia's new Bolshevik government had signed the Treaty of Brest-Litovsk with the Central Powers in March, taking it out of the war, the Germans began to release their Russian prisoners of war. Manpower was short in Germany, and they initially held back the able-bodied, but under the auspices of several Red Cross societies, they did release invalids at a rate of a few thousand a day, and it was probably these 'veritable walking dead men' who brought the flu to Russia.[2]

It made landfall in North Africa in May, and it seems to have circled Africa to reach Bombay (Mumbai) before the month was out. From India it travelled east. At some point, however, it may have met itself coming back, because there are reports of it in South East Asia in April.[3] Soon enough, it was in China. 'Queer epidemic sweeps North China', reported the *New York Times* on 1 June, and sipping their morning coffee, New Yorkers went on to learn that 20,000 cases had been recorded in the northern Chinese city of Tientsin (Tianjin) and 'thousands' more in Peking (Beijing). In the capital, 'the banks and silk stores have been closed down for several days in large sections, the police being unable to attend to their duties'. It erupted in Japan at the end of May, and by July it was in Australia. Then it seemed to recede.

This was the first wave of the pandemic, and it was relatively mild. Like seasonal flu, it caused disruption but no major panic. It did, however, create havoc in the European theatre of war, where it interfered significantly with military operations. Following the signing of Brest-Litovsk, as a result of which the Eastern Front ceased to exist, General Erich Ludendorff – architect of the German war effort – tried to pre-empt the arrival of American troops by launching an offensive on the one remaining major front, the Western Front. He saw this Kaiserschlacht, or Kaiser's Battle, as Germany's last chance of victory, and he had the newly liberated divisions from the east at his disposal. Despite initial successes, however, the offensive ultimately failed. Both sides were weakened by flu. As many as three-quarters of French troops fell sick that spring, and more than half the British force. Whole units were paralysed and makeshift military hospitals were bursting at their canvas seams. The situation at the front was dire. 'We were lying in the open air with just a ground sheet and a high fever,' recalled a British private, Donald Hodge, who survived. On the German side, 900,000 men were out of action.

Allied propagandists tried to turn the situation to their advantage. Leaflets were released over German positions informing them that if their own forces weren't capable of relieving them, the British would. The leaflets fluttered down over German cities too. When the British journalist Richard Collier was soliciting eye-witness testimony of the pandemic in the early 1970s, he received a letter from a German man named Fritz Roth who remembered, as a schoolboy, picking one up in Cologne. The civilian population of Germany had been close to starvation since the 'turnip winter' of 1916–17 – a failure of the potato crop that exacerbated underlying hardships caused by an Allied naval blockade. The wording of the pamphlet recalled by Roth translated approximately as 'Say your Our Father nicely, because in two months' time you will be ours; then you will get good meat and bacon, and then the flu will leave you alone.'

The flu did leave them alone that summer, generally speaking, though it wasn't completely absent from Europe. In late July, a Turkish Army officer named Mustafa Kemal was waylaid by it in Vienna on his way back to Constantinople (Istanbul). He had been inspecting German lines on the Western Front, and was unimpressed by what he had seen. Meeting his German ally, the kaiser, he told him bluntly that he expected the Central Powers to lose the war. (The Turkish officer recovered and went on to become the first leader of the Republic of Turkey, acquiring the surname Atatürk, or 'Father of the Turks'.)

In August the flu returned transformed. This was the second and most lethal wave of the pandemic, and again by consensus, it is described as having erupted in the second half of the month at three points around the Atlantic – Freetown in Sierra Leone, Boston in America, and Brest in France. It was as if it had been brewing in mid-ocean – perhaps in the Bermuda triangle – but it hadn't, of course. A British naval vessel brought it to Freetown, a ship steaming from Europe probably brought it to Boston, while Brest received it either with the continuing influx of AEF troops, or from French recruits arriving in the city for naval training. In fact, many French people thought at the time that it had come into France from Switzerland. The Swiss, meanwhile, thought it had entered their country from neighbouring Germany and Austria, despite their best efforts to impose quarantine measures at the borders. Though Switzerland was neutral in the war, under an agreement with the warring nations it received a constant flow of sick and injured POWs, whom it accommodated in internment camps in the Alps.

The forty-three-year-old psychoanalyst Carl Gustav Jung presided over such a camp, for interned British officers, in the picturesque mountain village of Château d'Œx. In the closing months of the war, camp discipline was relaxed, and the internees were permitted visitors. One of Jung's biographers recounts the following, possibly apocryphal tale: one day, Jung was talking to the visiting wife of a British officer. During the course of the conversation she told him that snakes in her dreams always meant illness, and that she had dreamed about a huge sea serpent. When, later, the flu broke out in the camp, Jung considered it proof that dreams could be prophetic.[4] The flu first appeared in Jung's camp in July. By 2 August there were reports of flu deaths among French soldiers returning home from Swiss camps.[5]

From Boston, Freetown and Brest the second wave spread outwards, helped on its way by troop movements. In early September, returning to New York from France on the troopship the SS *Leviathan*, Franklin Delano Roosevelt – a young assistant secretary of the navy at the time – developed symptoms and had to be carried ashore on a stretcher. Over the next two months, it spread from the north-eastern seaboard to the whole of North America, and thence through Central America to South America, which also received it from the sea (as did the Caribbean; Martinique was spared until the end of November, when it arrived, as so often, on a mailboat). South America, which had experienced no spring wave, reported its first cases after a British mail ship, the SS *Demerara*, docked at the northern Brazilian city of Recife on 16 September, with infection on board.

From Freetown, the flu spread along the coast of West Africa and inland via rivers and the colonial rail network. From railheads in the interior, infected individuals carried it by bicycle, canoe, camel and foot to the most remote communities. South Africa was wide open to disease, with its many ports and well-developed rail network, and the flu arrived in Cape Town in September, aboard two troopships that had called previously at Freetown. Between them, the *Jaroslav* and the *Veronej* were bringing home around 1,300 of the 21,000 members of the South African Native Labour Contingent who had served in France. Basic precautions were taken when the ships arrived, to isolate the infected, but not all the cases were identified, and some men were wrongly given the all-clear and allowed to board trains for home. From South Africa the flu spread rapidly through southern Africa up to the Zambezi River and beyond. The Horn of Africa received it in November, and Haile Selassie I, then regent of Abyssinia, reported that it killed 10,000 in the capital, Addis Ababa: 'But I, after I had fallen gravely ill, was spared from death by God's goodness.'[6]

In London, on 5 September, Sergei Diaghilev's Ballets Russes gave a performance of *Cleopatra* at the Coliseum Theatre. The great dancer and choreographer Léonide Massine was terrified of catching the flu. 'I wore nothing but a loincloth,' he later recalled. 'After my "death" I had to lie still for several minutes on the icy stage, while the cold penetrated to my bones . . . Nothing bad came of it, but the next day I learned that the policeman who always stood out front of the theatre, a great hulk of a man, had died of flu.'[7]

By the end of September the flu had spread through most of Europe, triggering another lull in military operations. The tubercular Franz Kafka caught it in Prague on 14 October and, confined to his sickbed, witnessed the collapse of the Austro-Hungarian Empire from his window. 'On the very first morning of Kafka's influenza,' wrote one of his biographers, 'the family was awakened by unusual sounds, the clank of weapons, and shouted orders. When they opened the curtains, they saw something alarming: entire platoons were appearing from the dark side streets in full marching order and beginning systematically to cordon off Altstädter Ring.'[8] The military had been mobilised to ward off the very real threat of revolution in the face of the disastrous supply situation and the movement, which was gathering strength, to declare an independent Czech state.

The modern history of Poland finds a sorry echo in its experience of the Spanish flu. By 1918, the country had been entirely erased from the map by its more powerful neighbours – Germany, Austro-Hungary and Russia – who had divided it up between them. The modern territory of Poland received the second wave of flu from all three partitioning states, and the disease fronts met at the River Vistula in Warsaw – the geographic heart of the country as it would be reconstituted later that year.[9] At the heart of Warsaw, Jan Steczkowski – the leader of the temporary government set up in the occupied Polish territories with the blessing of Germany and Austro-Hungary – fell ill.

The autumn wave swept in a broad diagonal from south-west to north-east across Russia – suggesting that those returning POWs were still a source of infection – but the flu was probably introduced to the vast Russian territory at different points around its borders within a matter of days or weeks. The London *Times* reported that it was in Petrograd (St Petersburg) as early as August, along with typhus, smallpox, meningitis and a surge of 'insanity'. The American historian Alfred Crosby noted that the AEF infected the White Sea port of Archangel, in the north, when it arrived there on 4 September to support anti-Bolshevik forces.[10] And before September was out, the newly formed People's Commissariat for Health in Moscow had received reports of it from all over the country.

The Russian civil war, the Trans-Siberian railroad and the struggle between Britain and Russia for control of Persia – the so-called 'Great Game' – all contributed to the flu's spread across northern Asia. It entered hapless Persia via several routes, but perhaps the most efficient, in terms of dissemination, was the north-eastern one via the shrine city of Mashed. It arrived in India in September, and by October it was back in China. In the last few days of that month, a dose of it forced the Japanese prime minister, Hara Takashi, to cancel an audience with the emperor (he survived, only to be assassinated three years later).

The epidemic was declared over in New York on 5 November, but it lingered on in war-torn Europe, drawn out by food and fuel shortages. As the weather turned cold, the French consul in Milan noted that housewives forced to queue for milk in freezing fog presented particularly easy pickings.[11] On her release from an English prison, the Irish patriot and suffragette Maud Gonne returned to Dublin to reclaim her house from the poet W. B. Yeats, to whom she had lent it. Yeats's heavily pregnant wife being ill with flu at the time, he turned Gonne away. The woman who had for so long been his muse, to whom he had addressed the lines 'Tread softly because you tread on my dreams', now bombarded him with hate mail, and Gonne's daughter recalled a terrible confrontation between the two, 'among the nurses and perambulators' of St Stephen's Green.[12]

Strangely, the flu saved at least one life that autumn – that of a young Hungarian physicist named Leo Szilard. He fell sick while training with his regiment at Kufstein in Austria, and was granted leave to return home to Budapest. There he was wheeled into a hospital ward that 'resembled a laundry', with wet sheets draped between the beds.[13] This humidity cure is unlikely to have contributed to his eventual recovery, but he was still in the hospital when he received a letter from his captain, informing him that the rest of his regiment had been killed at the Battle of Vittorio Veneto, on the Italian front. Szilard later moved to America and worked on the problem of nuclear fission. He would become known as one of the men behind the atom bomb.

On 9 November, the kaiser abdicated. On the 11th, the armistice was signed and celebrations broke out across the world, creating close to ideal conditions for a crowd disease. Thousands poured into the streets of Lima, Peru, triggering an explosion of flu in the days that followed. An armistice ball organised by the Red Cross in Nairobi had a similar effect in Kenya, while in London, the poet Ezra Pound wandered through the rainy streets 'to observe the effect of armistice on the populace' and came down with what he thought, at first, was a cold.[14]

By December 1918, most parts of the world were once again free of flu. Very few places on earth had been spared this murderous autumn wave, though there were examples: the continent of Antarctica; the tiny islands of St Helena in the South Atlantic Ocean and Marajó at the mouth of the Amazon River; the bigger island of Australia, a gleaming exception to the rule that humans could do little to protect themselves, since a strict maritime quarantine kept the flu out.

The Australian authorities lifted the quarantine early in 1919 – too soon, as it turned out, because it was then that the third wave struck. This wave was intermediate between the other two, in terms of virulence. More than 12,000 Australians died in the southern summer of 1918–19, after the virus finally gained a foothold there, but they weren't alone in having let down their guard. The third wave arrived while communities all over the world were still reeling from the second. It peaked in New York in the last week of January, and by the time it reached Paris the peace negotiations were underway. Delegates from a number of countries fell ill – proof, if proof were needed, that the virus transcended geopolitical boundaries.

Some have posited a fourth wave that struck northern countries in the winter of 1919–20, that may have claimed the lives of German political scientist Max Weber and, in Britain, Canadian physician William Osler – the man who coined the term 'old man's friend' for pneumonia – but this is usually excluded from the pandemic proper. Most people consider that the third wave – and hence the pandemic – was over in the northern hemisphere by May 1919. The southern hemisphere, however, had many more months of misery to come, since its pandemic was staggered in time with respect to the north.

Brazil experienced only one wave of flu – that of autumn 1918 – but in Chile a second wave struck a whole year later, while the most lethal wave to wash over the Peruvian capital was the third, in early 1920. The city of Iquitos lies deep in the Peruvian Amazon and, even today, can only be reached by river and air. Its isolation meant that it had only one brush with flu, late in 1918, but that same isolation, combined with poor access to healthcare, ensured that it was devastating. The death rate in Iquitos, at the time the centre of the Amazon rubber trade, was twice that recorded in Lima.[15]

This final surge of death was mirrored on the other side of the Pacific, in Japan. The 'Late Epidemic', as the Japanese called it (to distinguish it from the 'Early Epidemic' of autumn 1918), began in late 1919 and ran into 1920. On 18 March 1920, a Japanese farmer in Shōnai, 500 kilometres north of Tokyo, made the following note in his diary: 'Keishirō caught a cold and is coughing, so he went to visit the divine image of the Stop-Coughing Priest south of Kannonji village, to pray he would get over his cough.'[16] The entries either side of this one suggest that Keishirō, a member of the farmer's family, was suffering from Spanish flu. If so, his must have been among the last cases, because by then, the pandemic was over.

4

Like a thief in the night

The vast majority of those who caught the Spanish flu experienced nothing more than the symptoms of ordinary flu – sore throat, headache, fever. And as with ordinary flu, most of those who fell ill in the spring of 1918 recovered. There were rare cases in which the disease took a serious turn, and some of those unfortunates died, but though this was sad, it wasn't unexpected. The same thing happened every winter.

When the disease returned in August, however, there was no longer anything mundane about it. Now, what began as ordinary flu quickly graduated to something more sinister. The flu itself was worse, and it was also more likely to be complicated by pneumonia – in fact it was bacterial pneumonia that caused most deaths. Patients would soon be having trouble breathing. Two mahogany spots appeared over their cheekbones, and within a few hours that colour had flushed their faces from ear to ear – 'until it is hard', wrote one US Army doctor, 'to distinguish the colored men from the white'.[1]

Doctors labelled this chilling effect 'heliotrope cyanosis'. Like so many Bordeaux wine merchants, they tried to describe the colour in as precise terms as possible, believing that slight changes in tint were informative about the patient's prognosis. It was 'an intense dusky, reddish-plum' according to one doctor. As long as red was the dominant hue, there was room for optimism. But as soon as one 'would need to mix some heliotrope, or lavender, or mauvey-blue with red', the outlook was bleak indeed.[2]

Blue darkened to black. The black first appeared at the extremities – the hands and feet, including the nails – stole up the limbs and eventually infused the abdomen and torso. As long as you were conscious, therefore, you watched death enter at your fingertips and fill you up. When Blaise Cendrars called at 202 Boulevard Saint-Germain on 8 November, the concierge informed him that Mr and Mrs Apollinaire were both sick. Cendrars recounts that he bounded up the stairs and hammered at the door. Someone let him in. 'Apollinaire lay on his back,' he recalled. 'He was completely black.'[3]

Apollinaire died the next day. Once the black had set in, death came within days or hours. The distress of the bereaved was compounded by the look of the cadaver: not just the blackened face and hands, but the horribly distended chest. 'The body decomposed very quickly and the chest literally raised itself up, so that we had to push down my poor brother twice,' wrote one survivor. 'The coffin lid had to be shut at once.'[4] Inside the chest, at autopsy, pathologists found red, swollen lungs that were congested with haemorrhaged blood, and whose surfaces were covered in a watery pink lather. The flu's victims died by drowning, submerged in their own fluids.

Pregnant women who caught the flu suffered miscarriages and premature births with shocking frequency. People bled spontaneously from the nose and mouth. The aptly named *Leviathan* – one of the largest ships in the world at the time – left Hoboken, New Jersey for France on 29 September 1918, with 9,000 military personnel and the ship's crew aboard. The disease broke out as soon as it left harbour, and by the time it docked in Brest a week later, 2,000 men were sick and there had been around ninety fatalities on board. The scenes the ship's passengers witnessed during the voyage were Dantesque. The spaces between the bunks in the troop compartments were so narrow that nurses tending the sick couldn't avoid tracking blood between them. With the higher bunks unusable by the sick, semi-conscious men were laid out on the decks instead, which soon became slippery with blood and vomit. 'The conditions during the night cannot be visualised by anyone who had not actually seen them,' wrote one American soldier who made the crossing, and the 'groans and cries of the terrified added to the confusion of the applicants clamouring for treatment'.[5]

The whole constitution was affected. People said the Spanish flu had a smell, as of musty straw. 'I never smelt anything like it before or since,' recalled one nurse. 'It was awful, because there was poison in this virus.' Teeth fell out. Hair fell out. Some did not even show any signs before simply collapsing where they stood. Delirium was common. 'They became very excited and agitated,' wrote a doctor in Berlin. 'It was necessary to tie them to their beds to prevent them hurting themselves as they threw themselves about.' Another doctor in Paris observed that the delirium seemed to manifest itself, counter-intuitively, once the fever had broken. He described his patients' anxiety-provoking sensation that the end of the world was nigh, and their episodes of violent weeping.[6] There were reports of suicides – of patients leaping from hospital windows. Children died in tragic circumstances too, but while adults were described as 'leaping', children 'fell'. Near Lugano, Switzerland, a lawyer named Laghi cut his own throat with a razor, while a clerk who worked in the City of London didn't turn up for work one day. Instead, he took a train to Weymouth on the south coast of England and threw himself into the sea.[7]

People reported dizziness, insomnia, loss of hearing or smell, blurred vision. Flu can cause inflammation of the optic nerve, and one well-documented effect of that is impaired colour vision. Many patients remarked, on regaining consciousness, how washed out and dull the world appeared to them – as if those cyanosed faces had drained all the colour from it. 'Sitting in a long chair, near a window, it was in itself a melancholy wonder to see the colourless sunlight slanting on the snow, under a sky drained of its blue,' wrote American survivor Katherine Anne Porter, in her autobiographical short story *Pale Horse, Pale Rider.*[8]

The most terrifying thing of all, however, was the way it arrived: silently, without warning. It is a characteristic of flu that the period of high infectivity precedes the onset of symptoms. For at least a day, and sometimes longer, a person may appear to be well though they are infected – and infectious. In 1918, if you heard a neighbour or a relation coughing, or saw them fall down in front of you, you knew there was a good chance that you were already sick yourself. To quote one health officer in Bombay, the Spanish flu arrived 'like a thief in the night, its onset rapid, and insidious'.[9]

LOVE IN THE TIME OF FLU

When Pedro Nava arrived in Rio de Janeiro in August 1918, he was fifteen years old. He had come to live with his 'uncle' Antonio Ennes de Souza in the smart neighbourhood of Tijuca, in the north of the city. Ennes de Souza was actually a first cousin of Nava's father, José, but José had died in 1911, leaving his family in straitened circumstances and forcing them to leave the city. When the time came for Nava to study seriously, his mother sent him back to Rio, into the care of Uncle Ennes de Souza.

He was immediately entranced by his elegant and vivacious Rio relations, and by one visitor to the house in particular – a niece of 'Aunt' Eugenia named Nair Cardoso Sales Rodrigues. Describing the radiant Nair in his memoirs more than half a century later, he compared her to the Venus de Milo – with her 'lustrous complexion, her red petal-like lips, her wonderful hair' – and recalled with perfect clarity the night they both heard about the epidemic known as *espanhola*.[10]

It was late September, and as usual in the Ennes de Souza household, the papers were read aloud at the dinner table. They contained a report of 156 deaths on board the ship *La Plata*, which had sailed from Rio, heading for Europe, with a Brazilian medical mission aboard. The sickness had erupted two days out of Dakar on the west coast of Africa. But Africa was far away, and the boat was heading further still. What concern was it of theirs? Not reported that night – perhaps due to censorship, or perhaps because the press did not consider it sufficiently interesting – was the progress of a British mail ship, the *Demerara*, that had stopped in Dakar on its way out to Brazil. It had arrived in the northern Brazilian city of Recife on 16 September, with cases of flu on board, and was now heading south towards Rio.

After dinner, Nava went to sit by an open window with his aunt, whose back he obligingly scratched. Nair sat with them, and as she contemplated the tropical night, he contemplated her. When the clock struck midnight, they closed the window and left the room, but Nair paused to ask if they should be worried about the 'Spanish' sickness. Years later, Nava recalled the scene: 'We were standing, the three of us, in a corridor lined with Venetian mirrors in which our multiple reflections lost themselves in the infinity of two immense tunnels.' Eugenia told her there was nothing to worry about, and they parted for the night.

The *Demerara* entered Rio's port in the first week of October, without encountering any resistance. It may not have been the first infected ship to reach the capital, but from at least the time of its arrival, the flu began to spread through the poorer bairros or neighbourhoods of the city. On 12 October, a Saturday, a ball was held at the Club dos Diàrios, a favourite haunt of Rio's coffee barons and other powerbrokers. By the following week, many of the well-heeled guests had taken to their beds. So had the majority of Nava's fellow students. When he turned up at college on Monday morning, only eleven of the forty-six students in his year were present. By the end of the day, the college had closed indefinitely. Nava, who was told to go straight home and not to dawdle in the streets, arrived at his uncle's house, 16 Rua Major Ávila, to find that three members of the household had fallen ill since morning.

The city was totally unprepared for the tidal wave of sickness that now overtook it. Doctors kept up punishing schedules and returned home only to find more patients waiting for them. 'Agenor Porto told me that in order to have some rest, he had to lie hidden inside his landaulet [car] covered with canvas sacks.' Food, especially milk and eggs, ran short. Cariocas – as inhabitants of Rio are called – panicked, and the newspapers reported the deteriorating situation in the city. 'There was talk of attacks on bakeries, warehouses and bars by thieving mobs of ravenous, coughing convalescents . . . of chicken-stuffed jackfruits put aside for the privileged – the upper classes and those in government – being transported under guard before the eyes of a drooling population.'

Hunger invaded the house on Major Ávila. 'I got to know that drab companion,' Nava wrote. 'After one day on thickened fish stock, another on beer, wine, spirits and the dregs of olive oil, I can still remember the dawn of the third day. No breakfast, nothing to eat or drink.' Ennes de Souza, aged seventy-one, donned a wide-brimmed hat, took up a defensive stick and a wicker basket, and accompanied by his convalescent nephew Ernesto, 'pale with an unkempt beard', went out to see what he could procure for his famished family. 'After many hours they came back. Ernesto was carrying a bag full of Marie biscuits, some bacon and a tin of caviar, his uncle ten tins of condensed milk.' These precious stocks were strictly rationed by Aunt Eugenia, 'as if the house on Major Ávila were Géricault's raft after the shipwreck of the *Medusa*'.

An unexpected visitor turned up at the house: Nava's maternal grandfather. He was passing through, he said, from the neighbouring state of Minas Gerais – where the epidemic had barely got underway. He asked, of all things, to be shown the sights: Praia Vermelha, Sugarloaf Mountain. His grandson obliged, pausing in wonder at the sight of the Praça da República, the vast public space in the centre of the city, as empty as the moon. 'I would see it like that once again, forty-six years later, on 1 April 1964, but that was during the revolution.'

He recalled looking up at the sky and seeing a pumice-grey dome in which the sun appeared as a dirty yellow blot. 'The sunlight was like sand in the eyes. It hurt. The air we breathed was dry.' His intestines rumbled, his head ached. Falling asleep on the tram home, he had a nightmare that the staircase on which he was standing was falling away beneath him. He woke up shivering with a burning forehead. His grandfather delivered him home, where he gave himself up to the sickness. 'I kept on rolling down those stairs . . . The days of hallucination, sweat and shit had begun.'

At the time that Nava fell sick, Rio was the capital of a young republic. A military coup had brought the reign of Emperor Dom Pedro II to an end in 1889, and with the abolition of slavery the previous year, it had seen a massive influx of freed black and 'mulatto' slaves. The poorest moved into *cortiços* or slums in the city centre. The *cortiços* – the Portuguese word for 'beehives' – often lacked running water, sewers and proper ventilation. Living conditions were better there than in the *subúrbios*, the shanty towns expanding on the outskirts of the city, but the *cortiços* were more visible. White, middle-class cariocas saw them as parasitising the city proper. Aluísio Azevedo conveyed the fear that they inspired in his novel *O Cortiço*:

> For two years the slum grew from day to day, gaining strength and devouring newcomers. And next door, Miranda grew more and more alarmed and appalled by that brutal and exuberant world, that implacable jungle growing beneath his windows with roots thicker and more treacherous than serpents, undermining everything, threatening to break through the soil in his yard and shake his house to its very foundations.

When President Francisco de Paula Rodrigues Alves came to power in 1902, he launched an ambitious programme of urban renewal with the goal of turning Rio into a showcase of modern, republican civilisation. In his vision of the *cidade maravilhosa*, the marvellous city, there was no place for the *cortiços*, those nests of disease whose inhabitants, condemned by their biology, were 'locked into a vicious cycle of malnutrition and infection'.[11] They were razed and their inhabitants forced out. Six hundred homes were destroyed to make way for the magnificent Avenida Rio Branco, so that by the time the American travel writer

Harriet Chalmers Adams described the city in 1920, she could write that 'This portion of the city has been cooler ever since, as the breezes sweep through the wide avenue from waterfront to waterfront.'[12] But the easy mixing of the different classes that had once characterised Rio, their coming together in the seeking of pleasure – especially when it came to music and dancing – had gone. Now there was no area of carioca life in which rich and poor were not divided by an impenetrable gulf.

The president also set out to rid the city of infectious diseases, and in this he was aided by a doctor, Oswaldo Cruz, who in 1904, as head of the General Board of Public Health, had ordered a campaign of compulsory vaccination against smallpox. At the time, the vast majority of Brazilians had no grasp of germ theory. For many it was their first experience of state intervention in public health, hence something extraordinary, and poor cariocas rioted. The 'Vaccine Revolt', as it was called, was about more than one perceived violation, however. It was an expression of a broader class struggle over whom the city should serve – the Brazilian masses, or the European elite.[13]

A decade later, vaccination had been accepted by most Brazilians, but Cruz's unpopularity survived his death in 1917, and it was this legacy that shaped cariocas' response to the new disease threat in 1918. On 12 October, the day that the flu spread through the elegant guests at the Club dos Diàrios, the satirical magazine *Careta* (*Grimace*) expressed a fear that the authorities would exaggerate the danger posed by this mere *limpa-velhos* – killer of old people – to justify imposing a 'scientific dictatorship' and violating people's civil rights. The press portrayed the director of public health, Carlos Seidl, as a dithering bureaucrat, and politicians rubbished his talk of microbes travelling through the air, insisting instead that 'dust from Dakar could come this far'. The epidemic was even nicknamed 'Seidl's evil'. By the end of October, when half a million cariocas – more than half the population – were sick, there were still those among Rio's opinion-makers who doubted the disease was flu.[14]

By then, so many corpses lay unburied in the city that people began to fear they posed a sanitary risk. 'On my street,' recalled one carioca, 'you could see an ocean of corpses from the window. People would prop the feet of the dead up on the window ledges so that public assistance agencies would come to take them away. But the service was slow, and there came a time when the air grew filthy; the bodies began to swell and rot. Many began throwing corpses out on the streets.'[15]

'The chief constable was on the verge of despair when Jamanta, the famous Carnival reveller, came up with a solution,' wrote Nava. In daylight, Jamanta was José Luís Cordeiro, a journalist at the influential *Correio de Manhã* (*Morning Post*) – a newspaper that tended to take a disapproving stance when it came to Carnival. By night he was someone else, a prankster who 'had learnt to drive trams just for fun, as would suit a bohemian night-owl type like himself'.

As the *Correio* was apologising for its inability to meet its usual deadlines, due to sickness among its staff, Jamanta came into his own. 'He asked his bosses for a luggage tramcar and two second-class tramcars and swept the city from north to south.' With his grisly cargo he headed through the dark, empty streets for the São Francisco Xavier Cemetery in Cajú, in the north of Rio, where he unloaded his sinister caravan 'resembling the Ghost Train or Dracula's ship', before turning round to make another tour, 'even though the sun was already up'.

The bell at the gate of the Cajú cemetery would not stop tolling, driving those who lived nearby almost mad. Gravediggers couldn't dig fast enough; a thousand bodies awaited burial. To save time, they dug shallower graves. 'Sometimes the ditch was so shallow that a foot would suddenly bloom on the earth,' recalled the writer Nelson Rodrigues.[16] Amateur gravediggers were hired at advantageous prices. 'Then came the prisoners,' wrote Nava: 'Mayhem.' The convicts were enlisted to clear the backlog. Talk of horrors spread: of fingers and earlobes severed for jewels; of the lifting of the skirts of young girls; of necrophilia; of people buried alive. In the hospitals, it was said, at the same hour every night, 'midnight tea' was served to those who were beyond help, to speed them on their way to the 'holy house' – as coffin sellers euphemistically referred to the cemeteries.

Were the rumours true, or were they some kind of collective hallucination, one city's imagination let loose by fear? In the end, Nava concluded, it didn't matter, because the impact was the same. Terror transformed the city, which took on a post-apocalyptic aspect. Footballers played to empty stadia. The Avenida Rio Branco was deserted, and all nightlife ceased. If you caught a glimpse of a human being in the streets, it was fleeting. They were always running, black silhouettes against a blood-red sky, their faces contorted in a Munch-like scream. 'It just so happens that the memories of those who lived through those days are colourless,' wrote Nava, who may have experienced that strange distortion of colour perception reported by other patients too. 'No trace of early morning tints, shades of blue in the sky, twilight hues or moonlight silver. Everything appears covered in an ashen grey or a rotten red and brings back memories of rain and funeral rites, slime and catarrh.'

When he rose from his sickbed, thin and weak, he went to sit by a window that gave onto the street: 'In just one hour I saw three poorly attended funeral processions passing down Barão de Mesquita.' A servant told him that the girl he idolised, Nair, was now seriously ill. Struggling up the stairs, he peered around her door and was shocked by what he saw. Gone was the radiance, gone the lustrous complexion. Her lips were chapped and livid, her hair dull, her temples bony and concave. 'She was so changed it was as if she had turned into another person, as if some kind of demon were haunting her.'

Nair died on 1 November, All Saints' Day, by which time the epidemic was receding and life in Rio was returning to normal. It was pouring with rain. The hearse with its white curtains vanished 'as if in an aquarium', accompanied by Ernesto. When he returned that night, he told the others that the coffin had been lowered into a waterlogged grave. Five years later, when Aunt Eugenia went to retrieve Nair's bones, she found her 'completely incorrupt, only darkened like a mummy'. The gravedigger explained that her body had been preserved in its damp, anaerobic environment.

Nair was reburied in the earth, and after two years her clean bones were transferred to the family tomb. Nava's enduring image of her was of 'a marble bride' in a white dress, lying in a white coffin that the Venetian mirrors at 16 Rua Major Ávila reflected ad infinitum, her lips parted in a sad smile. 'She belonged to the past now, as distant as the Punic Wars, as the ancient Egyptian dynasties, as King Minos or the first men, errant and miserable.' From over fifty years' distance, the retired doctor bade her farewell: 'Sweet girl, may you rest in peace.'

PART THREE: *Manhu*, or What is it?

The Family, painted by Egon Schiele in October 1918

5

Disease eleven

When a new threat to life emerges, the first and most pressing concern is to name it. Once it has been named it can be spoken of. Solutions to it can be proposed and either adopted or rejected. Naming is therefore the first step to controlling the menace, even if all that is conferred with the name is the illusion of control. So there's a sense of urgency about naming; it must happen early. The trouble is that, in the early days of an outbreak, those observing the disease may not see the whole picture. They may misconstrue its nature or origin. This leads to all kinds of problems later on. The name first given to AIDS – gay-related immune deficiency – stigmatised the homosexual community. Swine flu, as we'll see, is transmitted by humans, not pigs, but some countries still banned pork imports after a 2009 outbreak. Alternatively, the disease may 'outgrow' its name. Ebola is named for the River Ebola in Central Africa, for example, but in 2014 it caused an epidemic in West Africa. Zika virus has travelled even further. Named for the forest in Uganda where it was first isolated in 1947, in 2017 it is a major threat in the Americas.

To try to prevent some of these problems, in 2015 the World Health Organization issued guidelines stipulating that disease names should not make reference to specific places, people, animals or food. They should not include words that engender fear, such as 'fatal' or 'unknown'. Instead, they should use generic descriptions of symptoms such as 'respiratory disease', combined with more specific qualifiers such as 'juvenile' or 'coastal', and the name of the disease-causing agent. When the need arises to distinguish between diseases that lay equal claim to these terms, this should be done using arbitrary labels such as one, two, three.

The WHO working group deliberated long and hard over this problem, which is not an easy one to solve. Take SARS, for example, the acronym for severe acute respiratory syndrome. It's hard to imagine how it could offend anyone, but it did. Some people in Hong Kong were unhappy about it – Hong Kong being one of the places affected by an outbreak of the disease in 2003 – because Hong Kong's official name includes the suffix SAR, for 'special administrative region'. On the other hand, names the current guidelines would rule out, such as monkey pox, arguably contain useful information about the disease's animal host and hence a potential source of infection. The working group considered naming diseases after Greek gods (Hippocrates would have been horrified), or alternating male and female names – the system used for hurricanes – but eventually rejected both options. They might have considered a numerical system that was adopted in China in the 1960s, in an attempt to prevent panic (diseases one to four were smallpox, cholera, plague and anthrax, respectively), but in the end they decided against too radical an overhaul. The current guidelines are designed to prevent the worst naming sins, while still leaving scientists room to be creative.[1]

These guidelines, of course, did not exist in 1918. Moreover, when influenza erupted that year, it did so more or less simultaneously all over the world, affecting populations that had embraced germ theory and others that had not. Those populations often had startlingly different concepts of disease per se. Since disease is broadly defined as the absence of health, whether or not you recognise a set of symptoms as a disease depends on your expectations of health. These might be very different if you live in a wealthy metropolis such as Sydney, or an Aboriginal community in the Australian Outback. The world was at war in 1918, and many governments had an incentive (more incentive than usual, let's say) to shift the blame for a devastating disease to other countries. Under such circumstances, that disease is likely to attract a kaleidoscope of different names, which is exactly what happened.

When the flu arrived in Spain in May, most Spanish people, like most people in general, assumed that it had come from beyond their own borders. In their case, they were right. It had been in America for two months already, and France for a matter of weeks at least. Spaniards didn't know that, however, because news of the flu was censored in the warring nations, to avoid damaging morale (French military doctors referred to it cryptically as *maladie onze*, 'disease eleven'). As late as 29 June, the Spanish inspector general of health, Martín Salazar, was able to announce to the Royal Academy of Medicine in Madrid that he had received no reports of a similar disease elsewhere in Europe. So who were Spaniards to blame? A popular song provided the answer. The hit show in Madrid at the time the flu arrived was *The Song of Forgetting*, an operetta based on the legend of Don Juan. It contained a catchy tune called 'The Soldier of Naples', so when a catchy disease appeared in their midst, Madrileños quickly dubbed it the 'Naples Soldier'.

Spain was neutral in the war, and its press was not censored. Local papers duly reported the havoc that the Naples Soldier left in its wake, and news of the disruption travelled abroad. In early June, Parisians who were ignorant of the ravages the flu had caused in the trenches of Flanders and Champagne learned that two-thirds of Madrileños had fallen ill in the space of three days. Not realising that it had been theirs longer than it had been Spain's, and with a little nudging from their governments, the French, British and Americans started calling it the 'Spanish flu'.

Not surprisingly, this label almost never appears in contemporary Spanish sources. Practically the only exception is when Spanish authors write to complain about it. 'Let it be stated that, as a good Spaniard, I protest this notion of the "Spanish fever",' railed a doctor named García Triviño in a Hispanic medical journal. Many in Spain saw the name as just the latest manifestation of the 'Black Legend', anti-Spanish propaganda that grew out of rivalry between the European empires in the sixteenth century, and that depicted the *conquistadors* as even more brutal than they were (they did bind and chain the Indians they subjugated, but they probably did not – as the legend claimed – feed Indian children to their dogs).[2]

Further from the theatre of war, people followed the time-honoured rules of epidemic nomenclature and blamed the obvious other. In Senegal it was the Brazilian flu and in Brazil the German flu, while the Danes thought it 'came from the south'. The Poles called it the Bolshevik disease, the Persians blamed the British, and the Japanese blamed their wrestlers: after it first broke out at a sumo tournament, they dubbed it 'sumo flu'.

Some names reflected a people's historic relationship with flu. In the minds of the British settlers of Southern Rhodesia (Zimbabwe), for example, flu was a relatively trivial disease, so officials labelled the new affliction 'influenza (*vera*)', adding the Latin word *vera*, meaning 'true', in an attempt to banish any doubts that this was the same disease. Following the same logic, but opting for a different solution, German doctors realised that people would need persuading that this new horror was the 'fashionable' disease of flu – darling of the worried well – so they called it 'pseudo-influenza'. In parts of the world that had witnessed the destructive potential of 'white man's diseases', however, the names often conveyed nothing at all about the identity of the disease. 'Man big daddy', 'big deadly era', myriad words meaning 'disaster' – they were expressions that had been applied before, to previous epidemics. They did not distinguish between smallpox, measles or influenza – or sometimes even famines or wars.

Some people reserved judgement. In Freetown, a newspaper suggested that the disease be called *manhu* until more was known about it. *Manhu*, a Hebrew word meaning 'what is it?', was what the Israelites asked each other when they saw a strange substance falling out of the sky as they passed through the Red Sea (from *manhu* comes *manna* – bread from heaven). Others named it commemoratively. The residents of Cape Coast, Ghana called it

Mowure Kodwo after a Mr Kodwo from the village of Mouri who was the first person to die of it in that area.[3] Across Africa, the disease was fixed for perpetuity in the names of age cohorts born around that time. Among the Igbo of Nigeria, for example, those born between 1919 and 1921 were known as *ogbo ifelunza*, the influenza age group. 'Ifelunza', an obvious corruption of 'influenza', became incorporated into the Igbo lexicon for the first time that autumn. Before that, they had had no word for the disease.

As time went on, and it transpired that there were not many local epidemics, but one global pandemic – it became necessary to agree on a single name. The one that was adopted was the one that was already being used by the most powerful nations on earth – the victors in the Great War. The pandemic became known as the Spanish flu – *ispanka, espanhola, la grippe espagnole, die Spanische Grippe* – and a historical wrong became set in stone.

6

The doctors' dilemma

The flu had been named; the foe had a face. But what did doctors *mean* by influenza in 1918? The most forward-thinking of them meant a constellation of symptoms that included cough, fever, aches and pains, that was caused by a bacterium named for its discoverer, Richard Pfeiffer. If a patient came to a doctor's consulting rooms complaining of feeling under the weather, the doctor might perform a clinical examination. He might take the patient's temperature, interrogate him about his symptoms, and look for the telltale mahogany spots over his cheekbones. This might be enough to convince him that the patient was suffering from flu, but if he was a rigorous type who wanted to be certain, he would take a sample of the patient's sputum (a polite word for coughed-up phlegm), grow its bacterial inhabitants on a nutritious gel, and then peer at them under a microscope. He knew what Pfeiffer's bacillus looked like – Pfeiffer himself had taken a photograph of it in the 1890s – and if he saw it, that would clinch the matter.

The trouble is, Pfeiffer's bacillus, though it is commonly found in the human throat, does not cause flu. In 1918, doctors found it in some of their cultures, but not in all. This violated the first of the great Robert Koch's 'postulates', the four criteria he had laid down for establishing that a particular microbe causes a particular disease: the microbe must be found in abundance in all organisms suffering from the disease, but should not be found in healthy organisms. Flu is, of course, caused by a virus. A virus is roughly twenty times smaller than a bacterium – too small to see under an optical microscope. Even if they had suspected that a virus caused flu, in other words, they had no way of detecting it. This was the doctors' dilemma in 1918: they did not know the cause of flu, so they could not diagnose it with certainty. And this dilemma bred others.

People were fairly easily convinced that the spring wave of the pandemic was influenza, but when the autumn wave erupted, there were serious doubts that it was the same disease. Even Americans and Europeans who had never seen a case of plague began to fear that that deadly disease had entered their midst. In hot countries it was mistaken for dengue fever, which also starts with a fever and headache. Cholera, which lends a blue tinge to the skin, was whispered, while a doctor at Constantinople's Hamidiye Children's Hospital claimed that it was worse than all of those, 'a disaster which isn't called plague but actually is more dangerous and more fatal than that'.[1]

Some doctors thought they were dealing with typhus, which starts out with flu-like symptoms – fever, headache, general malaise. Typhus has long been regarded as the disease of social collapse. It ravaged Napoleon's troops during their retreat from Moscow, and broke out in the Bergen-Belsen concentration camp in 1945 – probably killing the young diarist, Anne Frank. In 1918, when Russia was in the grip of a civil war, a doctor in Petrograd wrote that it 'follows Lenin's communism like the shadow follows the passer-by'.[2] That country experienced simultaneous epidemics of typhus and influenza, and Russian doctors were often at a loss to tell the two apart – at least until the patient broke out in the telltale typhus rash.

In Chile, doctors did not even consider the possibility of influenza. In 1918, Chilean intellectuals were gloomily convinced that their country was in a state of decline. The economy was faltering, labour disputes were on the rise, and there was a belief that the government was too much under the sway of foreign powers. When a new disease invaded the country, even though they had read reports of a flu epidemic in neighbouring states, a cabal of eminent Chilean doctors assumed that it was typhus. They blamed it on the poor and the workers, whom they referred to as *los culpables de la miseria* (makers of their own misery), because of the abject sanitary conditions in which they lived, and they acted accordingly.

Typhus is transmitted by lice, which means that it spreads much less easily than breath-borne flu. The Chilean doctors therefore saw no reason to ban mass gatherings. After ace pilot Lieutenant Dagoberto Godoy completed the first flight over the Andes in December 1918, ecstatic crowds greeted the hero in the streets of the capital, Santiago. Soon enough, the city's hospitals were turning away sick people for lack of space. Meanwhile, sanitary brigades went into battle against the imaginary typhus epidemic, invading poor people's houses and ordering them to strip, wash and shave body hair. In the cities of Parral and Concepción, they forcibly evicted thousands of workers and torched their homes – a strategy that probably exacerbated the epidemic, since it left crowds of homeless people exposed to each other and to the elements.

In 1919, while the country was still in the grip of the pandemic, a young woman entered the order of Discalced Carmelites in the township of Los Andes. Within a few months Teresa of Jesus – as the novice called herself – had fallen ill, and she died in April 1920 having taken her religious vows *in periculo mortis* (in danger of death). Teresa would later be canonised – English-speakers know her as Teresa of the Andes, Chile's patron saint. History books tell us that she died of typhus, but there is good reason to believe that she actually died of Spanish flu.[3]

The case of the Chinese province of Shansi (Shanxi) illustrates the doctors' dilemma most clearly, however, because it shows how difficult it was to identify a respiratory disease in a place where – as in many other parts of the world at that time – villages were isolated and hard to reach; where people were poor, often malnourished and suffering simultaneously from other diseases; and where they were opposed to 'foreign' medicine, creating conditions that were far from conducive to careful scientific work.

A TENTATIVE DIAGNOSIS

Shansi lies on China's frontier with Inner Mongolia. Surrounded on all sides by mountains and rivers, it is a landscape of precipices, ravines and rocky plateaus – the natural habitat of wolves and leopards. The Great Wall meanders through, a relic of attempts to keep out nomadic tribes and a reminder, along with the sandstorms that blow in from the Gobi Desert, of Shansi's position at the edge. In 1918, the people of the province lived in villages, but also in caves dug into cliffs. Their towns were fortified and protected by antique cannon. They were isolated by their geology, their geography and their history of conflict with outsiders, and all of this had left its mark. Fiercely proud of their ancient civilisation, they were considered conservative even by other conservative Chinese.

In 1911, revolution had overthrown the last imperial dynasty, the Qing, and ushered in a new republic. In the great cities of Peking, Shanghai and Tientsin, things were changing. The New Culture movement was challenging the rules by which Chinese society had organised itself for 4,000 years, and reserving particular scorn for traditional Chinese medicine. 'Our doctors do not understand science,' wrote Chen Tu-hsi (Chen Duxiu), one of New Culture's leaders, in 1915. 'They not only know nothing of human anatomy, but also know nothing of the analysis of medicines; as for bacterial poisoning and infections, they have not even heard of them.'[4] By 1918, however, these ideas had yet to percolate beyond the metropolises. Many in Shansi still recognised the Qing as their only legitimate rulers, and believed that illness was sent by demons and dragons in the form of evil winds. When disease scythed through them, as it did with dismal regularity, their first instinct was to appease the disgruntled spirits.

Though the revolution had given birth to a new republic, in reality the country had passed into the hands of rival provincial warlords. Yuan Shikai, the leader of the republic, had managed to keep these warlords more or less in check from Peking, but his death in 1916 ushered in a turbulent period during which they struggled to dominate one another. The governor of Shansi was a former revolutionary soldier named Yen Hsi-shan (Yan Xishan). Before the revolution, Yen had spent time in Japan which, unlike China, had embraced 'western' scientific ideas. On one occasion he had been admitted to a Japanese hospital, where he had encountered drugs and X-rays for the first time. He had seen with his own eyes how far his own country had fallen behind the rest of the world, and he had come to believe that Confucian values were toxic to it, hooks in its flesh anchoring it to the past. This 'enlightened' warlord was determined to remove those hooks, and to drag Shansi, bleeding if necessary, into the twentieth century.

Shansi's marginal position and natural ramparts meant that Governor Yen had less to fear from avaricious neighbours than other warlords, and he was able to divert his energies into an ambitious programme of reform. In 1917, he banned pigtails, the smoking of opium and foot-binding (which in Shansi meant wrappings to the knee, so that a woman's lower legs withered). Progressive societies were born – the Society for the Liberation of Feet, the Early Rising Society – and the local youth mobilised to enforce the new rules. Gangs of small girls chased transgressors through the streets shouting, 'Bad man, won't you be good!' All of Yen's reforms were unpopular, but the most unpopular of all were his attempts to control disease. Smallpox and TB were endemic to the region, while epidemics of plague, cholera and typhoid swept through as frequently as high winds in a hurricane belt. Despite their devastating impact, his efforts to quarantine the sick before an outbreak reached epidemic proportions were quietly circumvented. 'Few people were willing to be so unfilial as to turn their backs on a sick or dying relative,' wrote his biographer.[5] Filial piety, or respect for one's parents and elders, was a central pillar of the Confucian code.

In his battle to overcome this resistance, Yen came to lean heavily on the American missionaries who were the only providers of western-style medicine in Shansi. Many missionaries had been slaughtered in the Boxer Rebellion of 1900 – a violent uprising of the Chinese against western and Japanese influence (so-called because boxing was a ritual of the secret society that initiated it). Since then, however, a few brave souls had come to replace the victims. Yen admired them – men like Percy Watson, who ran the American hospital at Fenzhou (Fenyang), and Willoughby 'Dr Will' Hemingway, uncle of writer Ernest. At the first sign of a new epidemic, these men saddled up their mules and headed out to the often remote area where the first cases had been reported, to put into practice their modern ideas about hygiene, quarantine and cremation.

This is what they did again in October 1918, when the Spanish flu came to Shansi, and Governor Yen put the provincial police force at their disposal. 'Whole families were victims,' Watson wrote later. 'No family into which the disease entered, escaped with a mortality of less than 80 or 90 per cent, and the ones who escaped were mostly young children.' He added that 'it would be a conservative estimate to say, that there were not twenty Chinese in the entire province who did not believe that Chinese physicians could cure the disease'.[6] He was exaggerating, perhaps, but it was a telling comment: the population of Shansi was around 11 million at the time. Local people responded in their time-honoured way: 'They take the dragon god from the temple just north of our court, and with great noise, shouts and beating of drums, they go from home to home, hoping that the dragon through the noise can rid the city of these devils,' reported one missionary.[7]

Whether Watson's efforts were more effective is hard to know, but within three weeks the Spanish flu had receded. A few months of calm ensued, then on 7 January 1919 Governor Yen telegraphed Watson again, from the provincial capital Taiyuan, and asked him to investigate a new outbreak of disease in the mountains north-west of the city. Fenzhou was a hundred kilometres to the south-west of Taiyuan. With a team of experienced plague fighters, Watson set out, this time for a village called Wangchiaping, five days' trek from Fenzhou by pack mule. In this hill country, winters were cold. Villages were numerous but small, averaging three to four families each, and people worked the land to the tops of the mountains. When Watson arrived at the epicentre of the new outbreak, he discovered that it wasn't new at all. The first death had occurred a month earlier, on 12 December, but it hadn't been reported for more than three weeks. During that time, the disease had spread from Wangchiaping to nine other villages, carried either by relatives visiting the sick, or by those hired to bury the dead, or by a Chinese doctor who eventually succumbed to it himself.

As the missionaries moved from house to house, they came across scissors placed in doorways, apparently to ward off demons, 'or perchance to cut them in two'. In one of the nine affected villages, two orphaned children had been taken in by a couple who thought they had escaped the disease. 'When the children got sick it was at first decided to put them out in a temple and let them die there,' wrote Watson, 'but the man and his wife finally said they could not bear to do it, so the children were wrapped up in some bedding and left at one end of the *kang* bed* until they died the next day.'

Watson wasn't sure what the new disease was. He knew that the Wangchiaping district had been affected by the Spanish flu back in October, and he noted that the flu had been 'moderately severe' there, having been exacerbated by people crowding together on *kangs* to save on the *kaoliang* or millet stalks that constituted their only fuel. It was possible, however, that this new outbreak was a different disease: pneumonic plague. While all three varieties

* A *kang* is a large brick or clay heated platform used for sleeping on in northern Chinese houses.

of plague are caused by the bacterium *Yersinia pestis*, they present differently. The bubonic form is characterised by telltale 'buboes', when lymph nodes swell painfully; the septicaemic form arises from an infection of the blood; while the pneumonic or lung form is accompanied by chills and bloody sputum. Pneumonic plague is the most lethal of the three, and it may also be the most contagious, since it can be transmitted by the air.

The first death had occurred in Wangchiaping on 12 December, but Watson discovered that an elderly woman had died in another village in the same district on 28 November. This woman, he felt sure, had died of Spanish flu, because she had suffered profuse nosebleeds – a characteristic symptom of that disease, but not of pneumonic plague. In accordance with tradition, she had been left for ten days in an open coffin, in a courtyard, so that family members could pay their respects. 'It was in this courtyard that the first patient who died at Wangchiaping had been getting the lumber he was hauling before his sickness,' wrote Watson.

He had uncovered a direct link between the autumn epidemic and the one that had broken out in December. It seemed likely, therefore, that he was dealing with a new wave of Spanish flu – or the tail end of the previous one – but he hesitated to make that diagnosis. The later outbreak, in Wangchiaping, was both highly contagious and highly lethal. Around 80 per cent of those exposed to a living patient caught it, and none of them recovered – a pattern more characteristic of plague than flu, and different from that of the previous outbreak. The only way to settle the matter was to perform an autopsy, but Watson considered it unwise to do so. Mutilation of a corpse had been strictly forbidden under the Qing, and was still insulting to conservative Chinese. He also wanted to keep local people on side, so that they would comply with his quarantine measures.

He and his assistants did manage, cautiously, to obtain a few samples of sputum and lung tissue from victims using aspiration syringes. Emerging onto the hillside in plague suits, masks and goggles, Watson recalled, 'I fear we did little to lessen the Chinese idea of demons and ghosts.' Once they had got what they needed, a burial squad moved in, masked and gowned and armed with iron hooks with which to manipulate the cadavers. None of the doctor's microscopic examinations revealed the plague bacterium, though the traces of oedema or swelling that he saw in the aspirated lung tissue reminded him very much of the ravages of Spanish flu.

The 'new' epidemic was over by 25 January, two-thirds of the deaths having occurred before Watson arrived. Yen showed his gratitude by donating land to the American hospital in Fenzhou and decorating two Chinese members of its staff for their services in combating epidemics. His admiration for the missionaries was reciprocated. 'In the broadness of his interests and the dash and vigour of his personality he reminds somewhat of Theodore Roosevelt,' gushed one member of the Fenzhou mission.[8] Other epidemics would sweep the province, and as time went on, Watson measured the impact of the governor's modernisation efforts by his own practical yardstick: how many villages spontaneously organised their own quarantine at the first indication of an outbreak. Though he gave no details, he was clearly gratified by the results. By the 1930s, Shansi was regarded as a model province, and Yen a model governor.

7

The wrath of God

'In Bangkok,' wrote the British journalist, Richard Collier, 'the British Embassy's doctor, T. Heyward Heys, noted with dismay that almost all his prize roses had withered and died.' Owls came mysteriously to Paranhos da Beira, Portugal – a mountain village that had never known owls before – and hooted and screeched on every windowsill, while a faith healer in Montreal predicted a time of pestilence after the sky darkened in daytime, but no storm followed.[1]

Fear makes people vigilant. It impels them to notice things they might not otherwise have noticed; to pay attention to certain associations and ignore others; to remember prophecies they might previously have dismissed as absurd. Medieval chroniclers tell how, in the months before the Black Death arrived in Europe in the late 1340s, there were sightings of swarms of locusts, storms of hailstones 'of marvellous size', lizards and snakes raining from the sky. These frightening happenings testified to the corruption of the atmosphere that would soon bring an even greater evil in its wake: the plague.[2] They fit with medieval man's notion of miasma, or bad air, as the cause of disease. By 1918, germ theory had largely displaced miasma theory, but Galen's concept still lurked in the dark recesses of the human mind, and now it enjoyed a revival.

Some people claimed that the flu was caused by noxious vapours rising from the cadavers left behind on the killing fields. In Ireland, Kathleen Lynn, Sinn Féin's director of public health, told that organisation's council that the fever factory was 'in full working order in Flanders', and that 'the poisonous matter from millions of unburied bodies is constantly rising up into the air, which is blown all over the world by the winds'.[3] Others suspected human agency in the form of a secret programme of biowarfare conceived by one or other warring nation. This was less outlandish than it sounds. Biowarfare had a long and not very illustrious history. Possibly the first example of it occurred in the fourteenth century when, realising they had been infected with plague, Mongol forces laying siege to the Black Sea port of Kaffa (now Feodosia in Ukraine) hurled their dead over the city walls. Plague ripped through the city and the few who escaped fled west, taking it with them. Now, reaching for the aspirin packets manufactured by the German drug company Bayer, people in the Allied nations wondered if aspirin was *all* they contained. In Washington DC, meanwhile, newspapers reproduced comments made by Lieutenant Philip S. Doane, head of the health and sanitation section of the Emergency Fleet Corporation, in which he raised the spectre of German U-boats beaching in America and deliberately sowing the flu. 'The Germans have started epidemics in Europe, and there is no reason why they should be particularly gentle with America,' he was quoted as saying.[4]

These theories shrivelled and died, like Dr Heys' roses, as it became clear that soldiers on both sides of the front were dropping like flies. But other theories implicating an invisible hand took their place. How else could people explain the breathtaking cruelty of the disease? It became apparent very early on that, besides the elderly and the very young, it had a predilection for those in the prime of life – people in their twenties and thirties, especially men. Women seemed to be less susceptible, unless they had the misfortune to be pregnant, in which case, stripped of that invisible shield, they lost their babies and died in droves. The peak age of death in that middle group was twenty-eight, which meant that the disease was felling the pillars of families – including soldiers who had survived the war – and cutting the hearts out of communities. The Austrian artist Egon Schiele left a testament to that cruelty, in an unfinished painting he called *The Family*.

It portrays him, his wife Edith and their infant son, a family that would never exist because Edith died in October 1918, when she was six months pregnant with their first child. Schiele died three days later, having painted *The Family* in the interim. He was twenty-eight years old.

Another thing: how could you explain the randomness with which the disease selected its victims, if not as the work of a vengeful or vindictive force? Yes, the young and fit were in the firing line. But why was one village decimated, while a neighbouring one got away relatively unscathed? Why did one branch of a family survive, while a parallel one was snuffed out? In 1918 this apparent lottery was inexplicable, and it left people profoundly disturbed. Attempting to describe the feeling to Collier, a French doctor who was in the city of Lyons at the time, Ferréol Gavaudan, wrote that it was quite unlike the 'gut pangs' he had experienced at the front. This was 'a more diffuse anxiety, the sensation of some indefinable horror which had taken hold of the inhabitants of that town'.[5]

One of the most striking examples of this randomness took place in South Africa. The two great industrial centres in that country at the time were the gold mines of the Witwatersrand, or Rand, and the Kimberley diamond mines, dominated by the mighty De Beers Company. Between them these two economic powerhouses had driven the development of a railway system that had spread its black tentacles across southern Africa, keeping the ravenous maw supplied with cheap migrant labour. Trains set off from the port cities of Cape Town and Durban and penetrated deep into the rural interior, where they took their fill of African male youth before steaming a thousand kilometres north to disgorge their cargo into the mineral-rich heartlands.

The trains were spartan and became increasingly crowded as they stopped at each backwater on the two-day journey north (the Swazi word for train, *mbombela*, means 'a lot of people in a small space'). But if conditions were bad en route, they were only a taste of what was to come. At the mine compounds, men were allocated to overcrowded dormitories that consisted of concrete bunks built into the walls. Washing facilities were inadequate, food was scarce, and there was no privacy. Poorly lit and badly ventilated, these dormitories resounded with the miners' hacking coughs. Pneumonia was rife in Kimberley and on the Rand, because bacteria that caused it thrived in the cramped, humid underground spaces where the men worked. A lung that has been weakened by one infection is generally easier for another to invade. The two populations, in other words, were extremely vulnerable to a new respiratory disease – but on paper, at least, they were *equally* vulnerable.

Both the gold and diamond mines were affected early in the epidemic in South Africa, and because the companies that owned them kept track of their employees for accounting and legal purposes, we know what happened there. The flu reached the Rand first, by about a week, and the miners immediately knew they were confronting something other than the pneumonia epidemics that periodically tore through them, because this disease did not discriminate between new hands and old-timers. The vast majority of those who fell ill recovered, however, and the authorities breathed a sigh of relief. They then turned to watch with relative sangfroid as it advanced towards Kimberley. Their sangfroid turned to horror as the death rate in Kimberley quickly climbed to a staggering thirty-five times that recorded on the Rand. More than 2,500 diamond miners – almost a quarter of Kimberley's working population – died that autumn, and health officials could not explain why (they can now, as we'll see later).

In 1987, 43 per cent of Americans saw AIDS as divine punishment for immoral sexual behaviour.[6] In 1918, when a more mystical, pre-Darwinian era was still in living memory, and four years of war had worn down people's psychological defences, it was even easier for them to fall back on the belief that epidemics were acts of God. They looked for and found evidence to confirm them in that belief. A year earlier, the child visionaries of Fátima in Portugal, Jacinta and Francisco Marto and their cousin Lúcia

Santos, had claimed that the Virgin Mary had appeared to them on numerous occasions. Now, as the Marto siblings languished with flu, they reported a new visitation. 'Our Lady appeared to them and dissolved any possibility of a riddle with her simple statement that she would come for Francisco first and for Jacinta not long after that,' wrote their biographer. 'Their dry and fevered lips cracked under the strain of their smiles.' The children died, in the predetermined order, and their burial place became a site of Catholic pilgrimage.[7]

But if the Spanish flu was a punishment from God, what crime were people being punished for? There was no shortage of theories: the senseless war, of course, but also – depending on your position in society – degeneracy in the underclasses or the exploitation of indigenous peoples by their colonial masters. For some, however, it was retribution for something much more profound: people's decision to turn away, en masse, from the one true path. This was the case, for example, in one city in Spain.

A SIGNIFICANT VICTORY

The Spanish city of Zamora – known as *la bien cercada*, or well enclosed, due to its impressive fortifications – straddles the River Duero in the north-western region of Castile and León. Deeply religious, it is famous even today for its sombre processions of hooded, barefoot penitents in Holy Week. In 1914, when its citizens learned that they were about to receive a new bishop, the bells rang out for three days. The man himself arrived a few months later, stepping down from a specially chartered train to a railway station packed with well-wishers. Fireworks were let off, and a joyful crowd accompanied him to the cathedral where he took his oath of office. The church-sanctioned newspaper, *El Correo de Zamora*, promised obedience to the new bishop, and praised his eloquence and youth.

The bishop's name was Antonio Álvaro y Ballano, and at thirty-eight he already had a glittering career behind him. As a student at a seminary in Guadalajara, he had shone in every subject he had turned his hand to. At twenty-three he had taken up the chair in metaphysics, and after winning a hard-fought contest for the magistral canonry of Toledo, the most important archdiocese in Spain, he had come to the attention of Cardinal Sancha, Primate of Spain. He had been named a bishop in 1913, and prior to his arrival in Zamora, had held the post of prefect of studies at the seminary in Toledo.

In his inaugural letter to his new diocese, Álvaro y Ballano wrote that men should actively seek God and truth, which were the same thing, and expressed his surprise that science seemed to advance in step with a determination to turn away from God. The light of reason was weak, and 'modern societies mistake . . . contempt for God's law for progress'. He wrote of dark forces that wished to reject God 'or even annihilate him if that were possible'. The letter was peppered with scientific allusions, from Newton's law of universal gravitation to Ampère's experiments with compasses and electricity, although in his hands these became metaphors for describing the human soul's attraction to, or rejection of, God.[8]

The once-great Spanish Empire was at a low ebb. The Spanish–American War of 1898, *el desastre colonial*, had stripped it of its last imperial jewels – Puerto Rico, the Philippines, Guam, and the deepest cut of all, Cuba. It had contributed little to the great scientific and musical advances of the nineteenth century, and the golden age of Spanish literature was long behind it. Spanish society was still essentially agricultural, living conditions in some towns and cities were not so different from those that had prevailed in Europe at the time of the Black Death, and half the population was illiterate. 'The Madrid Spaniards are not accustomed to machinery or to industrialization,' observed the American writer and publisher Robert McAlmon. 'They have skyscrapers but they are rickety; they have elevators but they seldom work and then inspire one with fear of a crash; they have flush water-closets but even in the first-class hotels they are often clogged and dirty. The Spaniard is not modernized.'[9]

When the Naples Soldier returned to Spain in the autumn of 1918, it appeared first in the east of the country, but it soon followed the bishop along the train tracks to Zamora. September is a month of gatherings in Spain. The crops are harvested, the army takes on new recruits, and weddings and religious feasts are held – not to mention that most popular of Spanish pastimes, the bullfight. Young army recruits, some from distant provinces, converged on Zamora to take part in routine artillery exercises, and in the middle of the month, the *Correo* reported nonchalantly that 'There is cholera at the frontier, flu in Spain and in this tiny corner of the peninsula, *fiestas*.' Then the recruits began to fall ill.

Attempts to quarantine the sick soldiers in barracks on the site of the city's eleventh-century castle failed, and the number of civilian casualties began to rise. As it did so, the shortage of manpower began to interfere with the harvest, exacerbating pre-existing food restrictions. The press began to sound less sanguine. On 21 September, the *Heraldo de Zamora* – a newspaper that was nominally independent of the church – rued the unsanitary state of the city. Zamora resembled a 'pigsty' in which, shamefully, people still shared living space with animals, and many houses lacked their own lavatory or water supply. The paper repeated an old hobbyhorse, that the Moors had bequeathed to Spain an aversion to cleanliness. 'There are Spaniards who only use soap for washing their clothes,' it noted severely.

During the first wave of the pandemic, the country's inspector general of health, Martín Salazar, had lamented the inability of a bureaucratic and underfunded health system to prevent the disease from spreading. Though provincial health committees took their lead from his directorate, they had no powers of enforcement, and they quickly came up against what he described as the 'terrible ignorance' of the populace – the failure to grasp, for example, that an infected person on the move would transmit the disease. Now that the Naples Soldier had returned, one national newspaper, *El Liberal*, called for a sanitary dictatorship – a containment programme imposed from the top down – and as the epidemic wore on, the call was picked up and echoed by other papers.

In Zamora, the two local newspapers did their best to dispel public ignorance. They tried, for example, to explain the concept of contagion. The flu 'is always transmitted from a sick person to a healthy one', the *Correo* told its readers. 'It never develops spontaneously.' Local doctors weighed in, but not always helpfully. One Dr Luis Ibarra suggested in print that the disease was the result of a build-up of impurities in the blood due to sexual incontinence – a variation on the medieval idea that immoderate lechery could trigger a humoral imbalance. The papers published instructions from the provincial health committee for minimising infection – notably by avoiding crowded places. Yet they seem to have shown a mental block – at least to modern, secular eyes – when it came to the activities of the church. In a single issue of the *Correo*, an article approving the provincial governor's decision to prohibit large gatherings until further notice appeared alongside the times of upcoming Masses at the city's churches.

The papers also accused the authorities of playing down the gravity of the outbreak, and of not doing enough to protect people. Of national politicians, the *Correo* wrote, 'They have left us without an army, navy, bread or health . . . but nobody seems to resign or ask for resignation.' Local politicians, for their part, had long ignored calls to fund an infectious-diseases hospital, and were now ignoring recommendations from the provincial committee to impose stricter hygiene on the city. When a failure at a nearby hydroelectric dam led to a blackout, the *Correo* remarked with heavy irony that, despite the darkness, the hunger of Zamoranos and the filth in which they lived was plain for all to see. The night was densest inside the town hall, it quipped, which continued to plough money into bullfights but not into hygiene or food for a hungry population.

On 30 September, Bishop Álvaro y Ballano defied the health authorities by ordering a novena – evening prayers on nine consecutive days – in honour of St Rocco, the patron saint of plague and pestilence, because the evil that had befallen Zamoranos was 'due to our sins and ingratitude, for which the avenging arm of eternal justice has been brought down upon us'. On the first day of the novena, in the presence of the mayor and other notables, he dispensed Holy Communion to a large crowd at the Church of San Esteban. At another church, the congregation was asked to adore relics of St Rocco, which meant lining up to kiss them.

Also on 30 September, it was reported that Sister Dositea Andrés of the Servants of Mary had died while tending soldiers at the barracks. Sister Dositea was described as a 'virtuous and exemplary nun' who had accepted her martyrdom with equanimity and even enthusiasm, who had slept no more than four hours a day, and who had spent much of her time coaxing sick soldiers to eat. The Mother Superior of her order asked for a good turnout at her funeral, and the papers passed on her request. In accordance with tradition, readers were informed, the bishop would grant sixty days' indulgence to those who complied. Apparently the turnout was not as good as the Mother Superior had hoped, because the day after the funeral the *Correo* lambasted the citizenry for its ingratitude. The bishop, on the other hand, was satisfied with attendance at the novena, which he described as 'one of the most significant victories Catholicism has obtained'.

As the autumn wave neared its peak, fear and frustration threatened to spill over into unrest. Milk, which was being recommended by doctors to speed recovery, ran short and prices rocketed. Local journalists noticed that Zamoranos seemed to be dying in higher numbers than the residents of other provincial capitals, and they told their readers as much. They also returned again and again to the pitiful hygiene situation in the city. Residents simply threw their rubbish into the street, for example, and nobody seemed to care.

In October, the longed-for sanitary dictatorship came into effect. The authorities could now force businesses to close if they failed to meet sanitary requirements, and fine citizens who, for example, didn't keep their chickens cooped up. The provincial health committee threatened the city fathers with large fines for their laxity in recording flu deaths. But daily Masses continued to be held throughout that month – the worst of the epidemic – and the congregrations only grew as terrorised Zamoranos sought respite in the churches. The prayer *Pro tempore pestilentia*, which acknowledges that the affliction is God's will and that only His mercy will end it, echoed around their romanesque walls.

Despondency set in. There was a feeling that the horror would never cease, that the disease had become endemic. In a letter circulated on 20 October, Bishop Álvaro y Ballano wrote that science had proved itself impotent: 'Observing in their troubles that there is no protection or relief to be found on the earth, the people distance themselves, disenchanted, and turn their eyes instead toward heaven.' Four days later, a procession was held in honour of the Virgin of the Transit. People flooded into the city from the surrounding countryside, and the cathedral was packed. 'One word from the bishop was enough to fill the streets with people,' one paper reported. When the provincial authorities tried to use their new powers to enforce the prohibition on mass gatherings, the bishop accused them of interfering in church affairs.

As in other towns and villages, a decision was taken to stop ringing the church bells in eulogy of the dead, in case the constant tolling frightened people. But in other places, funeral processions had also been banned. Not in Zamora, where mourners continued to pass through the narrow streets as the din of the bells gave way to silence. Even in normal times, coffins – white ones for children – were a luxury beyond the means of most. Now, wood for coffins was hard for anyone to come by, and the bloated, blackened remains of the deceased were transported to their final resting place draped only in a shroud. In an echo of the ritual burning of incense to purify the altar, gunpowder was sprinkled in the streets and set alight. An approaching funeral cortège could thus be perceived only dimly through choking black smoke, mixed at times with the fog that rose from the Duero in those cool autumn days. 'The town must have looked as if it were on fire,' one historian observed.[10]

By mid-November, the worst was over. The bishop wrote to his flock attributing the passing of the epidemic to God's mercy. While expressing his sorrow for the lives that had been lost, he praised those who, through their attendance at the many novenas and Masses, had placated 'God's legitimate anger', and the priests who had lost their lives in the service of others. He also wrote that he felt comforted by the docility with which even the most lukewarm believers had received the last rites.[11]

The epidemic was not over when the bishop wrote his letter. There would be a reprise – milder than the autumn wave – the following spring. The journalists had been right: Zamora had suffered worse than any other Spanish city. But its residents do not seem to have held their bishop responsible. Perhaps it helped that they had grown up with the legend of Atilano, the first Bishop of Zamora, who in the tenth century had made a pilgrimage to the Holy Land to repent of his sins and free his city of plague. There are even those who defend Álvaro y Ballano, claiming that he did what he could to console his flock in the face of inertia at the town hall, the real problem being an ineffectual health system and poor education in matters of hygiene. Before 1919 was out, the city had awarded him the Cross of Beneficence, in recognition of his heroic efforts to end the suffering of its citizens during the epidemic, and he remained Bishop of Zamora until his death in 1927.

PART FOUR: The Survival Instinct

Royal S. Copeland and son, Washington DC, 1924

8

Chalking doors with crosses

Cordon sanitaire. Isolation. Quarantine. These are age-old concepts that human beings have been putting into practice since long before they understood the nature of the agents of contagion, long before they even considered epidemics to be acts of God. In fact, we may have had strategies for distancing ourselves from sources of infection since before we were strictly human.

Reading descriptions of the symptoms of the Spanish flu in these pages, you may have been aware of your own physical reaction of disgust. For a long time, scientists thought disgust was uniquely human, but they have come to regard it as a basic survival mechanism that occurs across the animal kingdom.[1] We avoid things we find disgusting, and such avoidance reactions have been observed in many species when contagion is a threat. The Caribbean spiny lobster, *Panulirus argus*, is highly sociable by nature, but it refuses to share a den with another lobster that is infected with a lethal virus. Chimpanzee troops steer clear of each other in the wild, not only to avoid unneighbourly disputes, but probably also to avoid contagion, while sick badgers in captivity have been observed anticipating a disgust response – or appearing to – by retreating to their tunnels and blocking them up with earth.

A sense of disgust, in this very basic sense of the term, may also be what drives animals to dispose hygienically of their dead. Honeybees scrupulously drag dead co-workers out of the hive, and elephants won't pass by a deceased one of their kind without covering it in branches and earth. Elephant-watcher Cynthia Moss tells how, after a cull in a park in Uganda, wardens collected the animals' lopped-off ears and feet in a shed, with the intention of selling them later for handbags and umbrella stands. One night, some elephants broke into the shed and buried the feet and ears.[2] The consensus among scholars is that humans started burying their dead systematically when they came together in the first settlements. Before that, they had left them exposed to the elements, and moved on.

Like chimps, human groups have probably been avoiding each other's germs for millennia, but as they became more sedentary, they were forced to come up with new strategies for keeping infection out. The much-feared sanitary cordon, in which a line is drawn around an infected area and no one is allowed out – sometimes on pain of death – is effective but brutal. In the seventeenth century, the English village of Eyam in Derbyshire erected a cordon around itself, once it knew it was infected with plague. By the time it was lifted, half the villagers were dead, but the infection had not spread. In the next century, the Habsburgs erected a cordon from the Danube to the Balkans, to keep infected easterners out of western Europe. Complete with watchtowers and checkpoints, it was patrolled by armed peasants who directed those suspected of infection to quarantine stations built along its length. Sanitary cordons fell out of favour in the twentieth century, but the concept was revived in 2014, during the Ebola epidemic in West Africa, when three affected countries erected one around the region where their borders met, believing this to be the source of the infection.

Another approach to containing disease is to forcibly isolate the sick or individuals suspected of infection in their own homes. This can work, but it is costly in terms of policing. More efficient, logistically speaking, is to round those individuals up in a designated space and keep them there for longer than the period of infectivity. Quarantine was invented by the Venetians in the fifteenth century, when they forced ships arriving from the Levant to sit at anchor for forty days – a *quarantena* – before they allowed those on board to land. The concept is much older, though. 'If the shiny spot on the skin is white but does not appear to be more than skin deep and the hair in it has not turned white, the priest is to isolate the affected person for seven days,' states the Bible (Leviticus 13:4–5). 'On the seventh day the priest is to examine them, and if he sees that the sore is unchanged and has not spread in the skin, he is to isolate them for another seven days.'

In the days before trains and planes, when most long-distance voyages were completed by sea, ports were the usual entry points for disease, and 'lazarettos' or quarantine hospitals were built either close to the docks or on offshore islands. They often resembled prisons, both in their architecture and in the way they treated their 'inmates', but by the nineteenth century enterprising merchants had realised that those inmates represented a captive market, and in some cities, they negotiated with the authorities to lay on restaurants, casinos and other forms of entertainment – all, of course, at elevated prices (today, many former lazarettos have become high-class hotels, so arguably not much has changed).

By the twentieth century, the problem of disease containment had become more complex. Infection didn't always arrive by sea, and the populations of the largest cities numbered in the millions. Their inhabitants not only did not know each other, beyond their own limited social networks, but they did not necessarily speak the same language or share the same beliefs either. In these modern cities, anti-infection measures had to be imposed from the top down, by a central authority. To pull this off, the authority required three things: the ability to identify cases in a timely fashion, and so determine the infection's direction of travel; an understanding of how the disease spread (by water? air? insect vector?), and hence the measures that were likely to block it; and some means of ensuring compliance with those measures.

When all three of these ingredients – which we'll describe in more detail in the following sections – were in place, containment could be extremely effective, but a hat-trick was rare. Often one or more were missing, meaning that an authority's efforts were only partially effective or ineffective. During the 1918 flu pandemic, all possible permutations were observed. We'll go on to explore two in particular – in New York City, and in the city of Mashed in Persia. In neither place was flu reportable at the beginning of the pandemic, but that is where the resemblance between them ends. Though more factors shaped their experiences of the flu besides their efforts to contain it, the contrast in the impact the epidemic had in the two cities was striking: the death rate from flu in Mashed was approximately ten times that in New York City.

Flagging up infection

The devastating plagues of the Middle Ages gave birth to the concept of disease surveillance – that is, the gathering of data on disease outbreaks so as to enable an appropriate and timely response, if not to the epidemic in progress then at least to the next one. To begin with, disease reporting was crude: diagnoses were vague, numbers approximate. Gradually, however, the data grew in volume and accuracy. Doctors started recording not only the numbers of sick and dead, but also who they were, where they lived, and when they first reported symptoms. They realised that by pooling and analysing these data, they could learn a great deal about where epidemics came from and how they spread. By the twentieth century, a number of countries had made disease reporting compulsory, and there was also recognition of the fact that infectious diseases don't respect borders. In 1907, European states set up the International Office of Public Hygiene in Paris, as a centralised repository of disease data, and to oversee international rules concerning the quarantining of ships.

In 1918, if a doctor diagnosed a reportable disease, he was obliged to notify local, state or national health authorities. The penalties for not doing so, though rarely enforced, included fines and revocation of his licence. Only diseases that were considered to pose a serious risk to public health were reportable, so that in the US, for example, smallpox, TB and cholera were reportable at the beginning of 1918, but influenza was not. Very few countries in the world that boasted well-organised disease-reporting systems required doctors to report flu at that time, which means, quite simply, that the Spanish flu took the world by surprise.

There were local reports of outbreaks, thanks mainly to the newspapers and to conscientious doctors who realised that this one was worse than most, but almost no central authority had an overview of the situation. Unable to connect the dots, they were ignorant of its date of arrival, point of entry, and speed and direction of travel. There was, in other words, no alarm system in place. The disease *was* made reportable, belatedly, but by the time the ancient instinct had been roused to batten down the hatches, it was too late: the disease was on the inside.

There were exceptions, but these owe their luck mainly to the happenstance of being islands, and remote ones at that. Iceland had a population of fewer than 100,000 at the time, and when the flu arrived in its midst, word quickly spread. Icelanders set up a roadblock on the main road leading to the north of the island, and posted a sentry at a place where an unbridged glacial river crossed a road, forming a natural barrier to the eastern part. Eventually, the authorities imposed a quarantine on incoming ships, and the combination of these measures helped keep more than a third of the Icelandic population flu-free.

Australia saw the epidemic coming from a long way off, both in time and space. Its authorities first heard about a flu epidemic in Europe in the northern hemisphere summer of 1918, and in September they became aware of the horrifying reports of the lethal second wave. Having watched it advance through Africa and Asia, they finally introduced quarantine procedures at all Australian ports on 18 October (New Zealand did not follow suit). When jubilant crowds gathered in Sydney's Martin Place to celebrate the armistice in November, therefore, they enjoyed the privilege – almost unique in the world – of having nothing to fear from the virus. Though the country did receive the third wave in early 1919, its losses would have been far greater had it let the autumn wave in.

The Philippines were not protected by their island status. When flu broke out there, it didn't occur to the occupying Americans that it might have come from outside, even though the first casualties were longshoremen toiling in the ports of Manila. They assumed its origins were indigenous – they called it by the local name for flu, *trancazo* – and made no attempt to protect the local population, which numbered 10 million. The only exception was the camp on the outskirts of Manila where Filipinos were being trained to join the US war effort, around which they created a quarantine zone. In some remote parts of the archipelago, 95 per cent of communities fell ill during the epidemic, and 80,000 Filipinos died.[3]

The starkly contrasting fates of American and Western Samoa – two neighbouring groups of islands in the South Pacific – show what happened when the authorities got the direction of travel right, and when they got it wrong. The American authorities who occupied American Samoa realised not only that the threat came from outside the territory, but also that indigenous Samoans were more vulnerable to the disease than white-skinned settlers, due to their history of isolation, and they deployed strict quarantine measures to keep it out. American Samoa got off scot-free, but Western Samoa, under the control of New Zealand, was not so lucky. After infection reached the islands via a steamer out of Auckland, local authorities made the same error as the occupiers of the Philippines, and assumed that it was of indigenous origin. One in four Western Samoans died in the ensuing tragedy which, as we'll see, would dramatically shape the islands' future.

Of course, for the most flagrant illustration of the global failure to report the Spanish flu, we need look no further than its name. The world thought it had come from Spain, when in fact only one country could legitimately accuse the Spanish of sending them the angel of death: Portugal. Injustice breeds injustice, and piqued at being made the world's scapegoat, the Spanish pointed the finger back at the Portuguese. Thousands of Spanish and Portuguese people provided temporary labour in France during the war, replacing French workers who had gone to fight, and though these labourers undoubtedly ferried the virus across borders, the Spanish singled out the Portuguese for blame. They set up sanitary cordons at railway stations, and sealed train wagons carrying Portuguese passengers so that they could have no contact with supposedly infection-free Spaniards in other wagons. At Medina del Campo, an important railway junction 150 kilometres north-west of Madrid, Portuguese travellers were sprayed with foul-smelling disinfectants and detained for up to eight hours. Those who protested were fined or even imprisoned. On 24 September 1918, much to the indignation of its neighbours, Spain closed both borders – a pointless move, since by then illness was already spreading through the castle barracks in Zamora. The Naples Soldier was back in the country.

Blocking the spread

An epidemic, like a forest fire, depends on 'fuel' – that is, individuals susceptible to infection. It grows exponentially from a few initial cases – the 'spark' – because those cases are surrounded by a vast pool of susceptible individuals. Over time, however, that pool shrinks as people either die or recover and acquire immunity. If you were to draw a graph of an epidemic, therefore, with 'number of new cases' on the vertical axis, and 'time' on the horizontal axis, you'd be looking at a normal distribution, or bell curve.

This is the classical form of an epidemic, though endless variations are possible – the curve's height or width may vary, for example, or it may have more than one peak. The basic form remains recognisable, which means that it can be described in mathematical terms. In the twenty-first century, the mathematical modelling of epidemics is highly sophisticated, but scientists had already begun to think that way in 1918. Two years earlier, in his 'theory of happenings', the British malaria expert and Nobel laureate Ronald Ross had come up with a set of differential equations that could help determine, at any given time, the proportion of a population that was infected, the proportion that was susceptible, and the rate of conversion between the two (with some diseases, infected individuals could return to the susceptible group on recovery). A happening, according to Ross's definition, was anything that spread through a population, be it a germ, a rumour or a fashion.

Ross's work, along with that of others, illustrated in hard numbers something that people had long understood instinctively – that a happening will begin to recede when the density of susceptible individuals falls below a certain threshold. An epidemic will run its course and vanish on its own, without intervention, but measures that reduce that density – collectively called 'social distancing' – can both bring it to an end sooner, and reduce the number of casualties. You can think of the area under the epidemic curve as reflecting the total amount of misery that it incurs. Now, picture the difference in size of that area when the curve is high and broad – that is, without intervention – and when it is low and narrow, with intervention. That is potentially the difference between an overwhelmed public health infrastructure, where patients can't get treated, doctors and nurses are pushed beyond exhaustion and dead bodies accumulate in morgues, and a functioning system that, though stretched to its limit, is still managing the flux of the sick.

In 1918, as soon as the flu had become reportable and the fact of the pandemic had been acknowledged, a raft of social distancing measures were put in place – at least in countries that had the resources to do so. Schools, theatres and places of worship were closed, the use of public transport systems was restricted and mass gatherings were banned. Quarantines were imposed at ports and railway stations, and patients were removed to hospitals, which set up isolation wards in order to separate them from non-infected patients. Public information campaigns advised people to use handkerchiefs when they sneezed and to wash their hands regularly; to avoid crowds, but to keep their windows open (because germs were known to breed in warm, humid conditions).

These were tried and tested measures, but others were more experimental. The Spanish flu was, to all intents and purposes, the first post-Pasteurian flu pandemic, since it was only well into the previous pandemic – the Russian flu of the 1890s – that Richard Pfeiffer had announced that he had identified the microbial cause of the disease. His model still prevailed in 1918, but it was, of course, wrong. With no diagnostic test available, and health experts disagreeing as to the agent of contagion – even, in some cases, the disease's identity – they found themselves caught on the horns of their own dilemma.

In some places, for example, the wearing of a layered gauze mask over the mouth was recommended – and in Japan this probably marked the beginning of the practice of mask-wearing to protect others from one's own germs – but health officials disagreed as to whether masks actually reduced transmission. They were divided over the use of disinfectant too. In late October 1918, well into the autumn wave – when metro stations and theatres across Paris were being doused in bleach – a journalist asked Émile Roux, director of the Pasteur Institute, no less, if disinfection was effective. The question took Roux by surprise. 'Absolutely useless,' he replied. 'Put twenty people in a disinfected room, insert one flu patient. If he sneezes, if a fleck of his nasal mucus or saliva reaches his neighbours, they will be contaminated despite the disinfected room.'[4]

It had long been assumed that school-age children represented ideal vectors of infection, because they are among the preferred victims of seasonal flu, they meet and mingle on a daily basis, and their snot control has a tendency to be suboptimal. The closing of schools was therefore a knee-jerk reaction, in case of a flu epidemic, and so it was in 1918. A couple of more thoughtful voices raised themselves against the clamour, however – and occasionally, as we'll see, even won the day. They belonged to observant individuals who had noticed two things: that school-age children were not the primary targets of this particular flu, and that even when they did fall sick, it wasn't clear *where* they had caught the disease – at home, at school, or somewhere in between. If it wasn't school, then closing the schools would neither protect the children nor stop the spread.

The most heated discussions of all, however, revolved around vaccination. Vaccination was older than germ theory – Edward Jenner had successfully vaccinated a boy against cowpox in 1796 – so it was undeniably possible to create an effective vaccine without knowing the identity of the microbe to which you were eliciting an immune response. Pasteur had, after all, created a vaccine against rabies without knowing that rabies was caused by a virus. In 1918, government laboratories produced large quantities of vaccines against Pfeiffer's bacillus and other bacteria thought to cause respiratory disease, and some of them actually seemed to save lives. Mostly, though, they had no effect: those who were vaccinated continued to fall ill and die.

We now know that the reason some of the vaccines worked is because they blocked the secondary bacterial infections that caused the pneumonia that killed so many patients. At the time, however, doctors interpreted the results according to their own pet theory of flu. Some pointed to the effective vaccines as proof that Pfeiffer's bacillus was the culprit. Others instinctively understood that the vaccines were dealing with the complication, not the underlying disease, the nature of which still eluded them. There were slanging matches, public disavowals. The American Medical Association advised its members not to put their faith in vaccines, and the press reported it all. The controversy was counter-productive, because the older measures – the ones that kept the sick and the healthy apart – worked, as long as people complied.

Getting people to comply

Quarantine and other disease-containment strategies place the interests of the collective over those of the individual. When the collective is very large, as we've said, those strategies have to be imposed in a top-down fashion. But mandating a central authority to act in the interests of the collective potentially creates two kinds of problems. First, the collective may have competing priorities – the need to make money, say, or the need to raise an army – and deny or water down the authority's powers of enforcement. And second, the rights of individuals risk getting trampled on, especially if the authority abuses the measures placed at its disposal.

The competing interests of the collective are the reason that historian Alfred Crosby, who told the story of the flu in America, argued that democracy was unhelpful in a pandemic. The demands of national security, a thriving economy and public health are rarely aligned, and elected representatives defending the first two undermine the third, simply by doing their job. In France, for example, powerful bodies including the Ministry of the Interior and the Academy of Medicine ordered the closure of theatres, cinemas, churches and markets, but this rarely happened, because prefects in the French departments didn't enforce the measures 'for fear of annoying the public'.[5] But a concentration of power at the top didn't guarantee effective containment either. In Japan, which was undergoing a transition from rule by a small group of oligarchs to a nascent democracy at the time, the authorities did not even consider closing public meeting places. A police officer in Tokyo observed that the authorities in Korea – then a Japanese colony – had banned all mass gatherings, even for worship. 'But we can't do this in Japan,' he sighed, without giving a reason.

Individuals also had cause to be wary in 1918. Throughout the last decades of the nineteenth century – that is, in very recent memory – public health campaigns had targeted marginalised groups, as eugenics and germ theory came together in a toxic mix. India is a case in point. The British colonial authorities had long taken a laissez-faire attitude to indigenous health in that country, believing it to be incorrigibly unhygienic, but when bubonic plague broke out in 1896, they realised the threat that deadly disease posed to their interests and went to the other extreme, imposing a brutal campaign to rout the infection. In the city of Pune, for example, the sick were isolated in hospitals, from which most never returned, while their relatives were segregated in 'health camps'. The floors of their houses were dug up, their personal effects were fumigated or burnt, and fire engines pumped such enormous quantities of carbolic acid into the buildings that one bacteriologist reported having to put up an umbrella before entering.[6]

Blinkered by their negative perception of the 'barefoot poor', the British authorities refused to believe – at least in the early days of the plague epidemic – that the disease was spread by rat fleas. If they had, they might have realised that a better strategy would have been to inspect imported merchandise rather than people, and to de-rat buildings rather than disinfect them. As for the Indians on the receiving end of these measures, they came to see hospitals as 'places of torture and places intended to provide material for experiments'.[7] Indeed, in 1897, the head of the Pune Plague Committee, Walter Charles Rand, was murdered by three local brothers, the Chapekars, who were hanged for their crime (today, a monument in the city honours them as freedom fighters).

Similar violations had taken place in other parts of the world. In Australia, a policy had been implemented to remove mixed-race Aboriginal children from their parents and place them with white families. The thinking was that 'pure' Aboriginals were doomed to extinction, but those whose blood was diluted with that of the 'superior' white races could potentially be saved by being assimilated into white society (this at a time when Aboriginals were dying in large numbers due to infectious diseases brought into their midst by white people). In Argentina, meanwhile, a programme had been launched to rid cities of people of African origin entirely, on the grounds that they posed a risk to the health of other citizens – a measure the Brazilian government considered but ultimately deemed unworkable because the vast majority of Brazilians were of African descent.

It was against this backdrop that, in 1918, health authorities once again announced the imposition of disease-containment measures. The pattern varied from country to country, but in general they were a mixture of mandatory and voluntary requirements. You were urged to use a handkerchief and to open your window at night, but nothing bad happened to you if you didn't. Vigilant police officers might stop you spitting in the street, and fine or even imprison you for a repeat offence, but if you violated the ban on mass gatherings by attending a political meeting or sporting event, you risked a band of them bursting in with batons and rudely breaking up the party. For breaching quarantine regulations, or a sanitary cordon, you could expect a very severe punishment indeed.

Many heeded the restrictions. This was a time, before civil rights movements, when authorities had more licence to intervene in private citizens' lives, and measures that would be perceived as invasive or intrusive today were more acceptable – especially in the patriotic atmosphere fostered by the war. In America, for example, it wasn't just conscientious objectors who were denigrated as 'slackers' in the autumn of 1918, but those who refused to comply with anti-contagion measures too.

Among those marginalised groups who had been targeted by such measures before, however, there were suspicions of another Trojan Horse, and many quietly rebelled. Vaccination programmes instigated in South Africa from November 1918 were widely boycotted. Both blacks and whites had a slippery grasp of germ theory, such that a contributor to the *Transkeian Gazette* was able to write that many thousands 'smile secretly when they are told that an inoculation dose contains so many millions of germs and humour the physician by pretending to believe'. But on top of that, blacks had to ask themselves why whites were suddenly so concerned about their health. There were rumours that white men were trying to kill them, with their long needles which – according to the rumour-mongers – they inserted into the jugular.

As time went on, fatigue set in even among those who had complied to begin with. Not only were the measures preventing them from going about their normal lives, but their efficacy appeared to be patchy at best. Role models forgot themselves. The mayor of San Francisco let his face mask dangle while watching a parade to celebrate the armistice. And the logic behind the restrictions was sometimes hard to follow. Father Bandeaux, a Catholic priest in New Orleans, protested the closing of churches in that city, when stores had been allowed to stay open. Such disparities, and the complaints they elicited, were duly reported in the newspapers.[8]

Newspapers were the main means of communicating with the public in 1918, and they played a critical role in shaping compliance – or the failure of it. They often took a lead in educating their readers about germ theory and passed on public health messages, but not without expressing opinions on them, and different newspapers expressed different opinions, sowing confusion. Their attitude, like that of doctors and the authorities, was paternalistic. Even in countries that weren't subject to wartime censorship, they rarely passed on information regarding the true scale of the pandemic, believing that the public couldn't be trusted with it. The concept of the 'mindless mob' was much more powerful then, and they were afraid of triggering panic. The masses were hard enough to 'steer' anyway – a widespread attitude that was summed up by the British newspaper the *Guardian* a few years later: 'But of what use is it to advise a modern urban population to avoid travelling on trains or trams, to ask the rising generation to abandon the pictures, or to warn the unemployed to take plenty of nourishing food and avoid worry?'[9]

The major Italian newspaper the *Corriere della Sera* took an original stance in reporting daily death tolls from flu, until civil authorities forced it to stop doing so on the grounds that it was stirring up anxiety among the citizenry.[10] The authorities don't seem to have realised that the paper's ensuing silence on the matter bred even greater anxiety. After all, people could see the exodus of dead bodies from their streets and villages. As time went on, and reporters, printers, truck drivers and newsboys fell sick, the news began to censor itself, and compliance dropped off even further. People drifted back to their churches, sought distraction in illicit race meetings, and left their masks at home. At that point the public health infrastructure – ambulances, hospitals, gravediggers – began to totter and collapse.

IMPERIAL METROPOLIS

New York in 1918 was the epitome of the atomised modern city. With a population of 5.6 million, it vied with London for the title of world's largest metropolis, and would overtake it within a few years. The reason for its rapid expansion was immigration. More than 20 million people arrived in the United States between 1880 and 1920, in search of a better life, and New York was their main port of entry. The vast majority came from southern and eastern Europe, and as with any immigrants far from home, it took them time to assimilate. New York in 1918 was many worlds within one world.

It was therefore a thoroughly modern challenge that faced the city's health commissioner, Royal S. Copeland, when the second wave of the epidemic declared itself in July: to elicit a collective response from a jumble of different communities who, though they overlapped in space, often had no common language and little shared identity. And that wasn't his only challenge. New York was the main embarkation point for troops heading for Europe, a role that ruled out the possibility of imposing an effective quarantine on the city.

Copeland was an eye surgeon and a homeopath – a less surprising combination in the days before homeopathy was labelled 'alternative' – and he had only been appointed to the commissioner's post in April of that year. A 'quintessentially optimistic, Bible-quoting, self-improving, platitude-sprouting American boy' from Michigan, he was seen as a practical man who got things done, yet that summer and early autumn, he dragged his feet.[11] The harbour authorities increased their surveillance of incoming ships from July, but when the thoroughly infected Norwegian vessel the *Bergensfjord* arrived on 12 August, and eleven of its passengers were taken to hospital in Brooklyn, they were not isolated. It was only on 17 September, by which time the epidemic was well underway, that influenza and pneumonia were made reportable in the city, and for the rest of that month Copeland played down the danger. By the time he officially acknowledged the epidemic, on 4 October, infected troopships including the *Leviathan* had long been ploughing backwards and forwards across the Atlantic, distributing their deadly cargo.

Copeland must have realised he was impotent where troop movements were concerned – President Woodrow Wilson had followed the advice of senior military officers, and overruled that of military doctors, in insisting that the transports continue – and he may have delayed declaring the epidemic so as not to impede them. Having declared it, however, he took three potentially life-saving decisions. First, he eliminated rush hour by staggering the opening times of factories, shops and cinemas. Second, he established a clearing-house system under which 150 emergency health centres were set up across the city to coordinate the care and reporting of the sick. And third and most controversially, he kept the schools open.[12]

Initially he had intended to close all public schools, as had happened in the neighbouring states of Massachusetts and New Jersey. But the pioneering head of the health department's child hygiene division, Josephine Baker, persuaded him not to. She argued that children would be easier to survey in school, and to treat should they show signs. They could be fed properly, which wasn't always the case at home, and used to transmit important public health information back to their families. 'I want to see if I can't keep the six-to-fifteen-year age group in this city away from danger of the "flu",' she told him. 'I don't know that I can do it, but I would awfully well like to have a chance.'[13] Copeland gave her that chance, and in doing so he brought bitter recriminations down on his head, including from the Red Cross and former health commissioners. But he and Baker would be vindicated: the flu was practically absent from school-age children that fall.

Copeland's campaign was repeatedly trumped by the demands of patriotism and the war effort. By 12 October, as the epidemic neared its peak, hospitals were seriously overcrowded, surgical wards were being turned into flu wards, and gymnasia and the city's first homeless shelter had been transformed to accommodate the overflow. But 12 October was Columbus Day, and to mark the occasion President Wilson led a parade of 25,000 down the 'Avenue of the Allies', as Fifth Avenue had been temporarily re-baptised.

Copeland also had to negotiate with local businessmen. Unlike health commissioners in other cities, he didn't close places of entertainment, though he did impose strict regulations on them – barring children, for example. When Charlie Chaplin's film *Shoulder Arms* – in which a tramp kidnaps the kaiser – was released on 20 October, the manager of the Strand Theatre, Harold Edel, praised his customers for their impressive turnout: 'We think it a most wonderful appreciation of *Shoulder Arms* that people should veritably take their lives in their hands to see it.'[14] Unfortunately, Edel died a week before his words were published, of Spanish flu.

In Copeland's favour, however, New York was practised in the art of public health campaigns, having declared war on TB – and particularly on the habit of spitting in public – twenty years earlier. By the end of September, the city was papered in advice on how to prevent and treat influenza. But the advice was printed in English, and it was only in the latter half of October, when the worst was already over, that boy scouts were sent scurrying through the tenements of Manhattan's Lower East Side to distribute pamphlets in other languages.

Of all the immigrant groups living in the city in 1918, the newest, poorest and fastest growing were the Italians. Around 4.5 million of them arrived in the four decades from 1880, and many of them never left. They gravitated to the 'Little Italies' – the Lower East Side, the area of Brooklyn around the navy dockyards, and East Harlem. They worked in factories and sweatshops, in construction or on the railroads, and they moved into crowded and substandard tenements, making New York the second most populous Italian city in the world after Naples.

Mainly *contadini* or peasants from rural southern Italy, these immigrants were unused to city life and particularly vulnerable to respiratory disease. We know this thanks to Antonio Stella, a doctor and respiratory disease specialist who was of Italian origin himself – having become naturalised in 1909 – and who went on to champion the Italian-American cause. When he wasn't seeing patients at the Italian Hospital on West 110th Street, or in his own consulting rooms, he was seeking them out in the parts of the city where they clustered. Sometimes he was accompanied by his younger brother, the artist Joseph Stella, who sketched what he saw and compared New York to 'an immense prison where the ambitions of Europe sicken and languish'.[15]

Long before the pandemic, Stella had noted the high rates of respiratory disease – especially TB – in the Little Italies, and the fact that the Italians had the highest mortality of any descendants of immigrant stock in the city. His forays into the notorious 'Lung Block' in Lower Manhattan, so-called because it was riddled with TB, convinced him that the problem was greatly under-reported – the true number of cases being perhaps twenty times that of which the city's health department was aware. 'Six months of life in the tenements are sufficient to turn the sturdy youth from Calabria, the brawny fisherman of Sicily, the robust women from Abruzzo and Basilicata, into the pale, flabby, undersized creatures we see, dragging along the streets of New York,' he wrote.[16]

Stella was acutely aware that disease could be used to further stigmatise already marginalised groups, and this was a time when much xenophobic feeling was directed against Italians. Considered unclean, slovenly and ruled by their passions, they were disproportionately blamed for crime, alcoholism, communism and a host of other social ills. For this respected and cultivated doctor, who collected antiques and damasks in his spare time, and who counted the rich and famous – including the celebrated tenor Enrico Caruso – among his patients, assimilation was the best protection against prejudice. His adopted city had demonstrated it repeatedly. Each new wave of immigration had been associated, not only with certain racial stereotypes, but also with specific diseases. In the 1830s, cholera was blamed on poor Irish immigrants. Towards the end of that century, TB became known as the 'Jewish disease' or the 'tailor's disease'. And when polio broke out in East Coast cities in 1916, the Italians were blamed. Visiting nurses disdained their practice of kissing the dead, and the Italians slammed their doors in the nurses' faces.

It was clear to Stella that the immigrants acquired most of their health problems in America – rather than bringing them with them, as nativists liked to say – and that the underlying problem was crowding in the tenements. In the worst cases he recorded, the density of human beings reached 120,000 per square kilometre, or nearly 500 per acre – higher than that of the most densely populated European cities at the time, and not far off that of Dharavi, the Mumbai slum considered one of the most densely populated places on earth today. In parts of East 13th Street, a Sicilian stronghold, he counted ten people, on average, living in a single room. But he was also aware that they exacerbated their own vulnerability with their backward ways. Many were illiterate and spoke no English. They were superstitious, clannish and mistrustful of authority. Their folk cures had undergone some modification, now that wolf bones were no longer available, but they found substitutes in the city's interstices, or cultivated them in window boxes. They continued to believe in witches and in the healing grace of the Virgin, and they spat to ward off the evil eye.

Most dangerously of all, to Stella's mind, these urban peasants believed in letting diseases run their course, in *pazienza* and what will be will be. They regarded doctors with the same suspicion they had priests and landowners in Italy, and considered hospital a place to die. Describing Manhattan's Bellevue Hospital (where Stella consulted) in his novel *The Fortunate Pilgrim* (1965), Mario Puzo wrote that 'The pious poor crossed themselves when they entered those gates.' In fact, Stella might have been the model for Puzo's Dr Barbato: 'Oh, he knew very well how they felt behind the respectful, honeyed *Signore Dottore this* and *Signore Dottore that*. He fed on their misfortunes; their pain was his profit; he came in their dire need and fear of death, demanding monies to succour them. In some primitive way they felt the art of healing to be magic, divine, not to be bought and sold.' The modern practice of paying for a doctor's services was foreign to them.

The main Italian-language daily newspaper in New York at that time was *Il Progresso Italo-Americano*. It sold close to 100,000 copies a day, and it was read as newspapers are, in communities with high levels of illiteracy: at the end of the working day, one semi-literate worker would read it many times, with great difficulty, then convey what he had understood to others, who would share and comment on it as they thronged the subway on their journey home. The writers of *Il Progresso* knew that their readers bound potato slices to their wrists to reduce fever, and kept their windows closed at night against evil spirits, and during the epidemic they took a stick-and-carrot approach to luring them away from such practices towards more 'orthodox' ones. The carrot was friendly advice: 'One should never kiss children on the mouth and should avoid kissing them as much as possible.' The stick was the law: 'Very strict orders have been issued against those who do not scrupulously follow hygienic measures or don't use a handkerchief when they expectorate. These infractions will be punished with both fines and jail time.'

Il Progresso was one of the few to voice approval when Copeland announced his decision to keep the schools open. Italian families tended to keep their children close – bringing them home for lunch, for example – but as the paper pointed out, children liberated from the classroom often went unsupervised in the streets, while in school teachers watched over them, and could spot the first signs. 'Moreover, in the schools hygiene and ventilation are better cared for than in many houses,' it added. In fact Copeland may have had the Italians in mind when he gave Baker's plan his blessing. Defending his father's decision years later, Copeland's son explained that, in one part of the Lower East Side, people 'were crowded ten to fifteen persons in two rooms and a bath. The tub was used to store coal. Hot water was non-existent and cold water was often lacking. People had to sleep in shifts. To close the schools would mean even greater exposure.'

Copeland himself caught the flu around the end of October, according to his son, but told no one and carried on managing the crisis. He declared the epidemic over on 5 November, though there would be a recrudescence in early 1919. When asked later why he thought the city had been affected less severely than other East Coast metropolises, he replied that New York was blessed with a solid foundation in public health, thanks to its twenty-year war on TB.[17] Most New Yorkers were familiar with the principles of hygiene, even if they didn't know how flu spread, and they were used to the authorities intervening in matters of their health. Another potentially protective factor, that he didn't mention, was the city's early and possibly prolonged spring wave of flu, that may have conferred some immunity on the population.[18]

The feared backlash against the Italians never came, and no other immigrant group was blamed for the flu either.[19] It has been suggested that the epidemic simply passed through too rapidly for people to start pointing fingers, but there may have been another reason too. Though everyone had been vulnerable to the flu, the Italians had been more vulnerable than most – and this was in the public record. The censors Copeland sent into the tenements when he made flu reportable – not only doctors, but also inspectors from non-medical agencies and laypeople – often arrived only in time to count the dead and arrange for their burial. Two weeks after he mobilised them, he gave them an extra task: to describe the sanitary conditions in which they found the patients. This added vivid detail to the sketchy picture of immigrant life the authorities had had to work with until then, and afforded better-off New Yorkers an unprecedented glimpse of the TB-ridden slums.

Il Progresso played its part in drawing attention to the Italian lot. In late October, it recounted the pitiful tale of Raffaele De

Simone, who had been unable to find an undertaker to furnish him with a coffin for his one-year-old baby. The little corpse had lain unburied at home for several days, until the desperate father, anxious for his four other children, had appealed for help: if any suitable box could be found, he would put his dead child in it, take it to the cemetery and even dig the grave himself if necessary (several days later, the paper reported that he had finally bought some wood and, in despair, fashioned a coffin with his own hands).

It probably also worked in the Italians' favour that the military had been badly affected (the US Army lost more men to flu than to combat, partly due to those deadly transports), and that many of the soldiers who died were of foreign birth. Some Italians had gone to Europe to fight even before America had entered the war, while an estimated 300,000 men of Italian origin joined the US Army.

Public funerals were banned in New York City during the pandemic, only spouses being allowed to accompany the coffin, but the authorities seem to have turned a blind eye on 27 October, when the funeral of Corporal Cesare Carella took place at the Church of Our Lady of Pompeii in Greenwich Village. Corporal Carella had survived the war only to die of Spanish flu, and large crowds gathered to watch his coffin pass on its way to the church. It was draped in the Italian flag, on which rested a bouquet of flowers and the distinctive wide-brimmed hat decorated with black capercaillie feathers that Carella had worn as a member of the Bersaglieri, a light-infantry unit of the Italian Army. American and Italian flags hung lowered from the windows along the route taken by the cortège, and the priest who addressed the packed church spoke, according to *Il Progresso*, 'as only an Italian priest can, who has faith and *la patria* in his heart'. Afterwards, the congregation accompanied the coffin all the way to Calvary Cemetery in Queens. There, unfortunately, there was a backlog of coffins waiting to be buried – Copeland discovered 200 of them when he visited two days later. The day after the health commissioner's visit, the mayor of New York, John Hylan, ordered seventy-five men to go to Calvary and clear the backlog.

Copeland may even have unwittingly pushed the Italians a step closer to assimilation. In September, a couple of days after making the flu a reportable disease – but before officially acknowledging the epidemic – he made hospitalisation compulsory for all flu patients living in shared accommodation. This included, of course, the crowded tenements. The reaction of the hospital-averse Italians is not recorded, except for one intriguing item in *Il Progresso*. The community, it reported on 25 September, had at last roused itself from its 'paralysing lethargy', and donated enthusiastically to a fund for a new Italian hospital in Brooklyn.

In fact, the sporadic cases of nativist prejudice that did occur in New York that autumn were directed mainly at those of German origin. In general, *Il Progresso* was punctilious about scotching wild rumours concerning the flu – including one that nurses and doctors found guilty of spreading flu germs among soldiers had been shot at dawn – but it couldn't resist passing on one story about a man who had allegedly given books to children outside a school on Long Island, telling them to scratch the pages to reveal images of President Wilson and other famous figures. His suspicions aroused, the headmaster had rounded the books up, noticed 'Made in Germany' printed on the back, and sent them off to be tested for flu germs (the results were awaited).

The flu also brought the Italians a new and powerful champion in Copeland. He now lent his support to reforms that Stella and other immigrant spokespeople had been advocating for years, declaring war on slum landlords and campaigning for better public housing. He argued that medical examinations of immigrants should take place *before* departure from the home country, to avoid exclusion on arrival, and lamented the waste of good farmers who were forced to scrape a living as 'poor city peddlers' in America. A year after the epidemic, the New York State legislature granted him $50,000 (around $900,000 in today's money) for a major study on the 'suppression and control of influenza and other diseases of the respiratory tract', and he went on to make sweeping changes to the city's public health structure, as part of which lectures were thenceforth organised in stores and factories in Italian and Yiddish.[20]

The city's first public housing project was initiated in 1934, on the Lower East Side. The mayor at the time was Fiorello La Guardia, the son of Italian immigrants and a former translator at the immigration depot on Ellis Island, whose first wife had died of TB aged twenty-six. 'Tear down the old, build up the new,' was how La Guardia presented the project at its unveiling in 1936. 'Down with rotten antiquated rat holes. Down with hovels, down with disease, down with firetraps, let in the sun, let in the sky, a new day is dawning.'

THE ILLNESS OF WINDS

Ahmad Qavam al-Saltaneh arrived in Mashed in January 1918, having made the ten-day journey across the desert from Tehran, probably in a horse-drawn diligence. I imagine him pausing on the Mount of Salutation – the place where pilgrims caught their first glimpse of the martyr's tomb – and gazing down on the golden dome glittering in the sunlight. He knew that a Herculean task lay ahead of him: the government had sent him to take charge of a city on the brink of anarchy.

Mashed at that time was the only city in Persia's vast north-eastern province of Khorasan. A sacred site for Shi'ite Muslims, it received the equivalent of its 70,000 inhabitants each year in pilgrims from all over the Shi'ite world. They came to pray at the shrine of Imam Reza, the eighth of twelve sacred imams considered by Shi'ites to be the spiritual heirs of the prophet Muhammad. But it was also a centre of the saffron and turquoise industries, known, too, for its beautiful carpets, and an important stop on the trade route from British India to the west, and from Persia to Russia.

The government, 900 kilometres away in Tehran, had little or no authority in Mashed, but Mashed was not immune from the political and economic crisis that had engulfed the rest of the country. For more than half a century, Persia had been a battleground for imperialist interests – the backdrop to the so-called 'Great Game', in which the British and Russians struggled for control of the huge area between the Caspian and Arabian Seas – and by 1918 its government was weak and almost bankrupt. Persia was, by then, a de facto protectorate.

A little over a decade earlier, in 1907, the British and Russians had signed a convention carving the country up into three zones – a northern Russian zone, a southern British zone, and a neutral zone in between – and this uneasy truce had held until war broke out in 1914. Persia declared itself neutral at the outset, but it didn't make much difference: it became a theatre of war by proxy. The British and Russians found themselves fighting on the same side against the Ottomans – who were threatening the country from the north-west – and their German allies. For the British, Persia was an essential buffer protecting the jewel in the crown, India, from their enemies, so when the tsarist Imperial Army collapsed in the wake of the Russian revolutions, and a power vacuum opened up in the northern zone, it worried them greatly. As soon as the Russians signed the Treaty of Brest-Litovsk, the British moved to occupy the whole of the east of the country. Mashed, which had always been a valuable listening post for them, became in the spring of 1918 an even more valuable military base.

By 1918, however, Mashed was not a comfortable place to be. It was a city under siege, controlled in all but name by tribes in the surrounding mountains who, long in the habit of robbing pilgrims as they approached the shrine on mule or horseback, were now brazenly sending raiding parties into Mashed itself. Pilgrims continued to converge on it, only now their numbers were being swollen by White Russian soldiers, many of whom had been wounded in fighting with the Bolsheviks further north. And there was famine. Two successive harvests had failed due to lack of rain, and the hunger had been exacerbated by the occupying armies' requisitioning of grain to feed their troops.[21]

Qavam set about restoring security to the city. He had a reputation as a deft negotiator, but he wasn't afraid to use force, if necessary, to get things done. He had some of the tribal chiefs arrested and placed in chains while they awaited punishment under sharia law. Public executions soon became a regular feature of Mashedi life. Some of the tribes were left in peace at the request of the British, who needed them for their troop levy, while with others he negotiated. 'The Governor General has satisfactorily settled the outstanding difficulty with the Hazara Chief Saiyid Haidar,' reported the British Consul General, Colonel Grey, a few months after Qavam's arrival. He had apparently persuaded Haidar to surrender 200 rifles in exchange for dropping all charges against him.[22]

The supply situation was harder to resolve. By the spring of 1918, the American minister to Persia, John Lawrence Caldwell, was reporting that Persians were eating grass and dead (rather than slaughtered) animals, even human flesh. The price of bread had quadrupled since 1916, though wages had not risen, and meat was no longer to be had in Mashed. The shrine was taking in abandoned babies, and people were lying down in the streets. Some sought sanctuary or *bast* in telegraph offices, an ancient custom in times of trouble, though their choice of refuge may have been inspired by a more recent belief that the wire would carry their pleas directly to the shah's palace in Tehran.

The famine was at its worst in June. By then, the British were feeding several thousand people a day, out of a courtyard at the consulate compound, though some have argued that the British relief effort was paltry by comparison with the stocks they had requisitioned.[23] Grey himself reported that during Ramadan, a well-known Mashedi preacher publicly criticised the British and threatened them with divine retribution. Typhus or typhoid or perhaps both were now raging in Mashed (there was diagnostic confusion over all the diseases present in the city at that time), and towards the end of June, cholera was reported further north, in the Russian city of Ashkhabad. Grey laid in supplies of serum from India and lamented the city's dismal sanitary situation: 'Nothing to be done regarding protection of the water supply.' In July it became clear that the next harvest would not fail, and the famine relief effort was eased back, but the British were still sufficiently concerned about cholera that they tried to discourage large numbers of people from making the traditional pilgrimage from what is now Pakistan to Mashed after the end of Ramadan.[24] They were still worrying about waterborne diseases when an airborne plague arrived in town – the Spanish flu.

It probably came in with a Russian soldier returning from Transcaspia, now Turkmenistan, along a rough-hewn road that wound through the Kopet Dag mountains in the north-eastern corner of the country. Its arrival in the third week of August coincided with that of a cold gale, causing local people to describe it as a disease of evil winds. Within a fortnight Grey was reporting that it had attacked every home and place of business, and that the platoons the British had levied – which were gathered in the city – were badly affected. The woeful inadequacy of the city's medical facilities now became apparent.

Apart from the British consulate's twelve-bed hospital and dispensary, there were two other conventional medical facilities for civilians in Mashed, both small in modern terms: one at the shrine, and one run by American missionaries. The shrine had had its own hospital since the nineteenth century (and some kind of medical facility long before that). It served mainly pilgrims, and occasionally miracles of healing were reported there. But by the time missionary doctor Rolla E. Hoffman arrived in Mashed in 1916, and visited it, he described it as 'a place where men went only to die; hardly a pane of glass in the whole place, wooden bedsteads without sheets or pillow cases, a dirt floor, no stove'.[25]

It might seem surprising that Presbyterian missionaries had dared enter such a holy Muslim site as Mashed, but years later another of them, William Miller, explained the impulse in startlingly simple terms: 'Since Meshed [*sic*] was an important center of Islamic devotion, it seemed incumbant [*sic*] on Christians to raise there the banner of Christ.'[26] The first to venture in, in 1894, was the Reverend Lewis Esselstyn. He caused a riot, and a kindly local smuggled him out again, but he was back in 1911, speaking Persian, and this time he managed to stay. When Hoffman joined him five years later, he quickly formed the opinion that it was only because of the medicine they provided that the Christians were tolerated there at all.

Mashed was still medieval in 1918, but its mud walls were crumbling. It was a city of graveyards where pilgrims who came to die had been buried on top of each other for centuries, and where, from time to time, old graves simply gave out, dissolving into the water supply. This took the form of man-made channels called *qanats*, that brought the water into the city from the nearby mountains. The water flowed uncovered down the middle of the main street – a permanent throng of pilgrims, merchants, camels and mules – and in the absence of a separate and enclosed sewage system, was easily contaminated. Germ theory had made its mark in Persia by 1918, but only in the literate 5 per cent. When it came to water, most people were guided by a religious prescription according to which water was clean if it was flowing, and if its volume exceeded one *korr* (350 litres). They therefore washed their pots and pans, their donkeys and themselves, very close to the open *qanats*.

The Tehran government had made repeated attempts to put in place a nationwide sanitary infrastructure, including a quarantine system for containing epidemics, but these had so far failed due to a lack of funds and because the British and Russians always managed to subvert them for their own political and commercial ends. For such an infrastructure to work, the country would have had to be united, and in 1918 it was far from that. Local attempts to improve hygiene in Mashed had also failed. When cholera broke out there in 1917, Qavam's predecessor had set up a sanitary committee that recommended long-term reforms such as moving the cemeteries outside the city walls, and introducing the reporting of contagious diseases, but none of them had been implemented.[27]

Because Mashed was a holy city, the shrine managers wielded great power – not only spiritually, but also financially, since the shrine owned large amounts of real estate. In 1918, Islamic thinking was still based on ninth-century teaching when it came to epidemics.[28] It accommodated the concept of contagion, but only up to a point. The general rule was that those inside a plague-stricken area should not flee, while those outside it, who were still healthy, should keep away. But there was also a fatalistic element to the prescriptions: plague was a martyrdom for believers and an agonising punishment for infidels. When sick, the vast majority of Persians turned to *hakims* or herb doctors, who followed two apparently complementary systems of medicine: the Galenic, and one that held that the Quran offered the best protection against disease. They might put an illness down to a humoral imbalance and recommend a change in diet, in line with the first; or they might identify the cause of the illness to be the sting of a jinn, and recommend strapping a prayer to the arm, in line with the second.

Esselstyn's missionaries had been run off their feet helping the British consulate distribute food during the famine. Both Hoffman and Esselstyn – who sometimes served as Hoffman's assistant, tucking his long red beard into his surgical gown – had caught typhus. Esselstyn died, and was buried in the Russian cemetery. Hoffman survived, only to come down with flu. It was at this point that Qavam made his first foray into public health. With the help of the British – and exploiting local fears of a Bolshevik invasion – he had by now taken control of most of the public institutions in the city, and revived the dormant sanitary committee. The committee in turn resurrected the recommendations it had made the previous year, during the cholera outbreak (it had barely had time to do anything else). They included a proposal that the burial of corpses inside the city be prohibited, at least for the duration of the epidemic, along with the bringing of corpses into Mashed from surrounding areas, and that a sanitary inspector oversee any burials that did take place inside the city walls.

Qavam wrote to the shrine managers on 18 September, asking them to implement the recommendations.[29] He was asking them to suspend centuries-old traditions, potentially even challenging sacred texts, and he must have anticipated the possibility of a rebuff, but his famous powers of persuasion saw him through. The shrine's chief administrator wrote back the same day, stating that, while he had not appreciated certain words and expressions the committee had used, which he had found insulting to the dignity of the shrine, he would nevertheless accede to the governor's request out of personal respect for him. He then wrote to his subordinates telling them what to do. Perhaps he had been impressed by the scale of the disaster himself, since he agreed that the committee's inspector be allowed to oversee burials, and even that the shrine underwrite his salary. Graves, he ordered, should be at least one metre deep. After the corpse had been placed inside, it should be covered with a thick layer of earth and lime, 'to eliminate the risk of noxious air rising from the corpse'. Anyone who failed to obey the new rules would be severely punished.

It was a breakthrough, of sorts, though not one that was likely to rein in an illness of winds – and certainly not at that late stage. The epidemic ran its natural course in Mashed. The worst was over by 21 September, by which time Khorasan and neighbouring Sistan provinces were thoroughly infected, and the flu was travelling west to Tehran at the speed of a 'prairie schooner' – the American nickname for a diligence. From Mashed, it rippled out with pilgrims, merchants and soldiers to the four corners of the country. By the end of September it was almost gone from the city, though it was still depleting outlying areas. At that point, life for Mashedis eased in one way and one way only: raids and attacks on pilgrims became rare. Qavam's policy of zero tolerance towards bandits may have begun to bite, but the hiatus was probably also an ominous sign of the havoc the flu had wrought in the mountains.

In a city with fewer than a hundred hospital beds, some 45,000 people, or two-thirds of the population, had come down with flu. An insight into the state of mind of the survivors – not only in

Mashed, but in Persia as a whole – is provided by the words of the city's chief astrologer, spoken at a public meeting towards the end of September. Astrologers were essentially mystics to whom Persians turned in times of crisis, and whose credibility was bolstered by the Islamic belief in predestiny. The chief astrologer relayed prophecies made a few days earlier by his counterpart in Tehran, to the effect that the British government would be annihilated the following year, 1920 would see the return to Persia of the current shah's father, who had been deposed in 1909, and 1921 the return of the Mahdi, the long-awaited Twelfth Imam, who would rid the world of evil.

October saw the beginning of the Shi'ite holy month, Muharram, which culminates in Ashura. The most sacred event in the Shi'ite calendar, Ashura commemorates the martyrdom of the Third Imam, Hussein. Several years later, the missionary William Miller would describe the Muharram processions as he witnessed them in Mashed: 'A group of men stripped to the waist was passing, beating their bare backs with chains,' he wrote. 'Then came the head-cutters, men who had made a vow to slash their foreheads with their swords until the blood streamed down their white gowns.'[30] The crowd looked on, lamenting loudly. Passion plays or *taziyehs* were performed. Muharram is a major event that absorbs all the city's energy for the month that it lasts, but in 1918 Grey wrote that it passed off quietly: 'Owing to recent sickness in the city the attendance at the processions was less than usual.'

Hoffman finally closed the American Hospital in December 1918, worn out from having run it single-handed throughout the crisis, and from his own bouts of typhus and flu. Before he took a well-earned rest, he managed to pen one last letter to the home church, asking for funds to support an expansion of the hospital and a second doctor. In the letter, he enthused about the potential for 'medical evangelism' in Mashed, to which all roads in the region led, and the possibility of offering pilgrims health in body *and* soul. The funds were duly granted.

Qavam survived the turbulence of General Reza Khan's British-backed coup in 1921, and finding favour with the new shah, went on to serve five terms as the country's prime minister. The shah eventually rebuilt Mashed on a rectilinear plan, linked it to Tehran by a modern road, and demolished its graveyards. Hoffman, who stayed on there until 1947, witnessed the transformation: 'The bones of centuries were shovelled into wheelbarrows and dumped into unmarked pits, the gravestones being used for street curbs and sidewalks.'[31]

9

The placebo effect

Much like today, when a person was sick in Europe or America in the late nineteenth century, he could go to a 'regular' doctor, or he could go to a homeopath, a naturopath, an osteopath or a faith healer – or he could hedge his bets and go to all five. The difference between then and now was that the regular doctor had no special status. There was nothing 'conventional' about his medicine or 'alternative' about theirs. His was just another cult among many medical cults. In the early twentieth century, the regulars fought off the competition from the 'irregulars'. They did this in Europe largely through increased state regulation of healthcare, and in America through a series of bitter legal battles, but the outcome in both places was the same: conventional medicine monopolised access to the masses. By 1918, it was indisputably mainstream.

When the Spanish flu broke out, therefore, it was to the regulars that most people in the industrialised world turned for treatment. What did the doctors have to offer? No effective vaccine, of course. Certainly no antiviral drugs – the first of those didn't enter the clinic until 1960 – and no antibiotics either, for the treatment of those opportunistic bacterial infections. They wouldn't become available until after the Second World War. Faced with wheezing, blue-faced patients, they felt they had to do something, and the approach they adopted was polypragmatism, or polypharmacy: they threw the medicine cabinet at the problem.

What was in a conventional doctor's medicine cabinet in 1918? It was still an era of 'concoctions, plant extracts, and other unproven treatments'.[1] Clinical drug development was in its infancy, and though some drugs had been tested in animals or humans, many had not. When human trials had been conducted, they tended to be small. The elaborate and very expensive drug trials that we read about today, with their 'blinded' experimenters and placebo controls, were unheard of. Legislation to ensure that medicines were pure and unadulterated was recent, in those countries that even had it. There was no real understanding of how a drug's active components interacted with living tissues, or the conditions that turned a medicine into a poison, and even when there was, most practitioners were unaware of it; it didn't form part of their training.

One of the first phials they reached for was the one containing aspirin, the 'wonder drug' that was known to reduce fever and kill pain. They prescribed it in such large quantities that in 2009, a doctor named Karen Starko put forward the troubling theory that aspirin poisoning might have contributed to the deaths of a sizeable proportion of the flu's victims. Very high doses of aspirin can cause the lungs to fill with fluid, a fact of which doctors were ignorant in 1918, and we know that they were routinely prescribing twice the maximum dose that is considered safe today. The aspirin-poisoning theory is contentious, however. Other scientists have pointed out that the drug wouldn't have been widely available in many countries – most Indians would not have had access to it, for example – so while it may have aggravated the situation in America and other wealthy countries, it is unlikely to have contributed to the global death toll in any significant way.[2]

It is nevertheless possible that many of those who suffered from Spanish flu also had to contend with the effects of overdosing of the substances that doctors gave them to try to ameliorate their symptoms. Quinine, for example, was a known treatment for malaria and other 'bilious fevers of a paludic nature'.[3] There was no evidence that it worked for flu, yet it was prescribed in large doses. 'To the symptoms of the disease now had to be added those caused by the panacea: buzzing in the ear, vertigo, hearing loss, bloody urine and vomiting,' wrote Pedro Nava in Brazil.

Though rare, disturbed colour vision can be a side effect of taking a lot of quinine – meaning that this drug may have exacerbated the sensation that some flu victims had, of coming to in a pallid, washed-out world.

Arsenic preparations were popular, for their tonic, painkilling action, as was camphor oil for treating shortness of breath. Digitalis and strychnine were supposed to stimulate the circulation, Epsom salts and castor oil were prescribed as purgatives, and various drugs derived from iodine for 'internal disinfection'. When none of these things worked, doctors fell back on older techniques. Having observed that some patients seemed to take a turn for the better following a gushing nosebleed, menstruation, even – traumatically – miscarriage, some revived the ancient practice of bloodletting, or medicinal bleeding. Physicians of the Hippocratic and Galenic traditions thought that this cleansed the blood of impurities, and in 1918 it was commented upon that the blood that emerged from flu patients was unusually thick and dark. The practice provoked a certain amount of scepticism, however. 'Although this resource did not relieve or cure anyone, it brought comfort to the patient and the family,' wrote one Spanish doctor.[4]

Even more controversial was alcohol – especially in those states where the prohibition movement was gaining force, and it couldn't be obtained without a prescription. Some doctors claimed that alcohol in small doses had a stimulant effect, while others urged complete abstinence. Vendors seized on these slim pickings to trumpet the medicinal properties of their wares. Afraid of provoking a different kind of epidemic, health officials in the Swiss canton of Vaud circulated a memo that urged doctors to 'vigorously oppose the idea taking root that alcohol in high doses protects against influenza' – even if, those same officials allowed, it could be useful when the patient was feverish and unable to feed himself. Some doctors claimed that inhaling cigarette smoke killed the germ, and people naturally cherry-picked the advice that suited them. The Swiss-born architect known as Le Corbusier retreated to his rooms in Paris and sipped cognac and smoked through the worst of the pandemic, while cogitating on how to revolutionise the way people lived (though he hadn't even a diploma in architecture).

Some enterprising 'experimentalists' suggested new prophylactics or therapies based on their observations. While treating patients at the New Idria mercury mine in San Benito County, California, physician Valentine McGillycuddy noticed that none of the men who operated the furnace where the metal was extracted from its ore had contracted the flu. This, he surmised, was due to mercury's antiseptic properties, or else to the fact that mercury vapour stimulates the salivary glands (McGillycuddy turns up again later in Alaska – we'll meet him there). French military doctors observed, apparently independently, that when the flu invaded an army clinic for venereal disease, all the patients succumbed except the syphilitics, and they wondered if it was these patients' daily mercury injections that protected them. A Viennese doctor went so far as to conduct a small trial. Since none of his twenty-one flu patients died following mercury treatment, he concluded that it was an effective therapy for flu.[5] Unfortunately, as many syphilitic patients discovered, mercury is also toxic. The symptoms of mercury poisoning include loss of coordination and a sensation of ants crawling beneath the skin. The therapy, in this case, was arguably worse than the disease.

It wasn't so difficult, in the circumstances, for the manufacturers of patent medicines to tap into a newly receptive audience, and make small fortunes selling their dubious products over the counter. Their tonics and elixirs – Dr Kilmer's Swamp-Root was a famous formula of the time, in America – were generally plant-based too, and often claimed heritage among ancient people's recipes. These days, research into ethnic groups' use of native plants, ethnobotany, is a respected field in its own right, and pharmaceutical companies search for potential new blockbuster drugs in those indigenous pharmacopoeias. But in 1918, patent medicines were relatively unregulated and there was rarely any evidence that they worked. The regulars – arguably standing on shaky ground themselves – accused their makers of quackery.

Those who had no truck with either turned instead to the irregulars. Having suffered the ill effects of overdosing with conventional medicines, the prospect of a 'nature cure' or an extremely diluted homeopathic compound might well have appealed to them. Alternatively they put their faith in home remedies: mustard poultices, sugar lumps soaked in kerosene, infusions brewed according to old family recipes, fires of aromatic plants lit twice daily in front of the house (to clear the miasma).

Beyond the industrialised world, people sought out their traditional healers – sometimes after western-style doctors, sometimes before. Ayurveda in India, like *kanpo* in Japan – ancient forms of treatment that make use of herbs – were trusted and cheap alternatives to western medicine which, even if they had confidence in it, was often not available. Witch doctors in the hills of India moulded human figures out of flour and water and waved them over the sick to lure out the evil spirits. In China, besides parading the figures of dragon kings through their towns, people went to public baths to sweat out the evil winds, smoked opium and took *yin qiao san* – a powdered mix of honeysuckle and forsythia that had been developed under the Qing for 'winter sickness'.

Most of these 'cures' were no more effective than placebos. The placebo effect is a manifestation of the power of positive thinking. It derives from a person's expectation that a drug or other intervention will heal them, and it can be extremely effective in itself. According to some estimates, 35–40 per cent of all medical prescriptions today are not much more than placebos.[6] The interesting thing about a placebo is how sensitive it is to the trust that is established between a patient and his doctor. If a patient loses confidence in his doctor, or if he perceives that the doctor has lost respect for him, the beneficial effects of the placebo shrink – and that shrinkage doesn't necessarily stop at zero. It can enter negative space, giving rise to a harmful or 'nocebo' effect.

Some of the therapies prescribed in 1918 are described as having aggravated the symptoms. They may have actually done so, for biochemical reasons, or they may have been acting as nocebos –

and this applied to western and traditional remedies alike. The term 'nocebo' did not enter the conventional medical lexicon until the 1960s, yet some healers may have instinctively grasped the concept. There are reports of shamans fleeing when they saw that their ministrations were having no effect. Perhaps they feared for their lives, or perhaps they understood that they risked doing more harm than good. Western doctors, adhering to a different code of conduct, stayed at their posts, trying treatment after treatment in the hope of finding one that worked. In fact, there were really only two things that any physician could do to improve his patient's chances of survival: ensure that he didn't become dehydrated, and that he was carefully nursed.

People expected more, of course – in part because more had been promised. Disappointed, many turned to higher authorities. Muslims sought sanctuary in mosques, while Jewish communities all over the world performed an archaic ritual known as a 'black wedding' – the best description of which comes from Odessa, Russia, and will be presented in the next section. In the melting pot that was New York City, this produced the intriguing juxtaposition, on the Lower East Side alone, of Italian immigrants pleading for *la grazia* – the Virgin's healing grace – while their neighbours, Jews from eastern Europe, witnessed the nuptials of two of their number among the gravestones of Mount Hebron Cemetery. When God Himself proved impotent, people gave up, and like sick badgers, immured themselves in their homes.

BLACK RITES

When the first wave of Spanish flu struck Russia in May 1918, it went virtually unnoticed in most of the country, but not in Odessa, where a doctor named Vyacheslav Stefansky recorded 119 cases at the Old City Hospital.

The surprise is not that this wave went unnoticed elsewhere, but that the Odessans noticed it. In 1918, Russia was in the grip of a civil war following the revolutions of the previous year.

Odessa is now in Ukraine, but in 1918 it was the third most important city in the Russian Empire after Moscow and Petrograd, and a key battleground in that war in southern Russia. Odessans, who are known in Russia for their mischievous sense of humour, liked to compare their city to a prostitute who goes to bed with one client and wakes up with another. In 1918 alone, it passed from the Bolsheviks to the Germans and Austrians (under the terms of Brest-Litovsk), to Ukrainian nationalists and, finally, to the French and their White Russian allies.

Odessa did not witness the violence known as the Red Terror that ruptured the northern cities – though it did not entirely escape the killings, torture and repression instigated by the Bolshevik secret police, the Cheka – but it did experience a breakdown in the bureaucratic underpinning of life, resulting in food and fuel shortages and a security vacuum into which local crime lords sharply stepped. One nicknamed Misha Yaponchik – the model for Isaac Babel's Jewish gangster Benya Krik, in his 1921 *Odessa Tales* – took control of the streets with a gang that allegedly consisted of 20,000 bandits, pimps and prostitutes, and like a latter-day Robin Hood proceeded to terrorise the better-off.

Odessa differed from the two northern cities in other ways, too. It was warm, pleasure-loving, cosmopolitan and open to the west. It had a large Jewish contingent – a third of its 500,000-strong population according to official figures, more than half according to unofficial ones. And it was more advanced in the understanding and surveillance of infectious diseases. This Black Sea port, known as the 'Russian Marseilles', had for centuries provided a stop on the route by which silks and spices from the east were transported westwards to Constantinople and beyond. It had always been vulnerable to pathogens arriving by sea, and almost since Catherine the Great gave it city status in 1794, had operated a quarantine system. Quarantine had rarely kept disease out entirely, however, as the city's many plague cemeteries testified. The most visible of these, a plague mound known as Chumka, still stands on its outskirts.

It was logical, therefore, that in 1886, Ilya Mechnikov should choose Odessa as the site of Russia's first plague control facility – the Odessa Bacteriological Station. This was set up as a result of Pasteur's development – with Émile Roux – of a rabies vaccine, and it had the mission of producing and perfecting vaccines of all kinds. Within its first six months of existence, it administered anti-rabies shots to 326 individuals from all over Russia, Romania and Turkey, who had been bitten by rabid animals. Mechnikov soon fell out with his Russian colleagues, however. Unlike them, he was a bench scientist, not a medic, and he found it difficult to impose his authority on them. When he moved to Paris two years later – desolate at having to leave his beloved Russia – he bequeathed the station to his capable (and medically qualified) assistant, Yakov Bardakh.

Under Bardakh's direction the station carried out important research into anthrax, typhoid, cholera, malaria and TB. When he introduced the inspection of drinking water and the tests revealed typhoid bacteria, the sanitarians responsible for the city's water supply attacked him, refusing to believe that the disease was waterborne. He was later vindicated, but when poor people started lining up outside the station for treatment, it was too much. Odessa had long been regarded as a hotbed of revolutionary dissent, and the authorities placed the station under police surveillance.

Perhaps on account of those bedraggled lines, perhaps because he had experimented with deadly diseases, or perhaps because he was Jewish, Bardakh was removed from his post in 1891. Russian law barred Jews from the headships of certain institutions, and strict quotas governed how many of them could enter education and employment. Some Jews took Russian names to circumvent these restrictions, but not Bardakh. 'I am a Jew' he wrote proudly on every official document that required him to state his ethnic origin. Mechnikov lamented his departure: 'Science lost a gifted worker.' But when Pasteur offered Bardakh a post in Paris, he refused, preferring to stay and serve his country.[7]

The directorship of the station passed to his student, Stefansky, and Bardakh entered private practice. The authorities could not stop his reputation from growing, however. He saw patients at the city's Jewish Hospital and in his own home. Though he came from humble stock – he was the son of a Jewish scholar and teacher – his wife, Henrietta, was the daughter of a banker, and they received a constant stream of visitors in the large, oak-panelled dining room of their home on Lev Tolstoy Street, where Henrietta served tea from a samovar. So many people arrived at Odessa train station asking for Bardakh that the coach drivers all knew his address by heart. He taught bacteriology at the city's university – the first such courses in Russia – and he launched the Odessan tradition of lecturing the public on science. Large audiences came to hear him talk about the origins of plague and Pasteur's discoveries, and he regularly kept them glued to their seats until midnight. By 1918, Bardakh was the most famous doctor in southern Russia, and his name was mentioned with respect in capitals further west, too.

After the spike of cases of *ispanka* in May, the numbers fell back in June and July. *Carpe diem* was the motto of Odessans that summer, and the universe seemed to conspire with them to help them forget their troubles. In June, finding himself in the occupied city, an Austrian officer commented on their vivacity, carelessness and gusto, and in the same month Vera Kholodnaya arrived in town. The twenty-four-year-old actress was the undisputed queen of the Russian screen, as famous for her grey, hypnotic gaze as for her roles as beautiful, betrayed women. She had come as part of a mass exodus of artists from Moscow and Petrograd, where political and economic chaos were stifling the film industry. Adoring crowds greeted her, and her gaze rested on them all through August as her latest film, *The Woman Who Invented Love*, played in cinemas. Rumours about the star's suspected involvement in the underground provided another marvellous distraction. In Kiev, according to the writer Konstantin Paustovsky, it was said that she 'had recruited her own army like Joan of Arc and, riding a white charger at the head of her victorious troops, had entered the town of Priluki and proclaimed herself Empress of Ukraine'.[8]

The illusion of glamour and romance was shattered on 31 August by a series of powerful explosions at a munitions depot in the poor suburb of Bugaevka. The alleged work of White Russians, who wanted to prevent the planned handover of artillery shells to the Germans and Austrians, the explosions destroyed most of the buildings in a seven-kilometre strip that included granaries, a sugar factory and hundreds of homes. Reuters reported that the death toll was 'limited', but thousands were left without food or shelter, exposed to the elements, and in the first days of September, cases of *ispanka* began to trickle and then flood into the Jewish Hospital.

By now, besides *ispanka*, the city was also dealing with cholera – which had arrived on an Austrian troopship in August – and the countrywide epidemic of typhus. The German and Austrian occupiers were no more interested in addressing the city's health issues than they were in addressing its crime problem. Their sole objective was to requisition the region's grain stores, to send back to their starving compatriots, and they ensured the minimum of security to get that job done. As a result, this city that was so practised in the art of quarantine – and that had tracked the flu since May – had no strategy in place for containing it. Cafés and theatres remained open, crowds filled them seeking oblivion or at least temporary distraction, and Yaponchik's gangsters raided their homes while they were out – or even while they were in.

Bardakh did what he could. Chairing a packed meeting of the Society of Odessa Doctors, he told them that their individual efforts to fight flu among the poor and the working class were of little use in the absence of a city-wide containment programme. Echoing Copeland in New York, he said that closing schools was only advisable if there was evidence to show that children were contracting the disease at school, rather than at home, and he pointed out that the expression 'infection through the air' had been widely misunderstood. He knew that in the poorer neighbourhoods of the city, homes were dark, damp and overcrowded – havens for germs – and that even the well-off and educated were suspicious of fresh air. It must be hammered home, he told the assembled doctors, that though one should avoid people who were coughing, fresh air was vital for preserving good health.

Given that there was no ban on public meetings in the city, Bardakh seems to have decided that he might as well continue with his programme of public education, perhaps with the aim of stimulating a 'bottom-up' response to the epidemic. That autumn, he and other leading doctors spoke in cinemas, theatres, synagogues, the famous Pryvoz food market – even the city's opera house, during an interval in *Faust*. He reassured his listeners that *ispanka* was not some new and terrifying form of plague, as many feared, but a virulent form of influenza, and that one *could* protect oneself from it – notably by ventilating one's home. Some did not want to hear such rational explanations, and on 1 October, Odessa bore witness to a black wedding.

A *shvartze khasene*, to give it its Yiddish name, is an ancient Jewish ritual for warding off lethal epidemics, that involves marrying two people in a cemetery. According to tradition, the bride and groom must be chosen from among the most unfortunate in society, 'from amongst the most frightful cripples, degraded paupers, and lamentable ne'er-do-wells as were in the district', as the nineteenth-century Odessan writer, Mendele Mocher Sforim, explained in his fictional description of such a wedding.

Following a wave of black weddings in Kiev and other cities, a group of Odessan merchants had got together in September, when both the cholera and *ispanka* epidemics were waxing, and decided to organise their own. Some in the Jewish community disapproved strongly of what they considered a pagan and even blasphemous practice, but the city rabbi gave his blessing, and so did the mayor, who deemed it no threat to public order. Scouts were dispatched to the Jewish cemeteries to search for two candidates among the beggars of alms who haunted those places, and a suitably colourful and dishevelled bride and groom were chosen. Once they had agreed to be married in their 'workplace', the merchants set about raising funds to pay for the festivities.

Many thousands gathered to witness the ceremony, which took place at three o'clock in the afternoon in the First Jewish Cemetery. Afterwards, a procession headed for the city centre accompanied by musicians. When it arrived at the hall where the reception was to be held, the sheer number of people pressing in to get a look at the newlyweds prevented them from climbing down from their carriage. Eventually the crowd fell back and the couple were able to enter the hall, where their nuptials were celebrated with a feast and they were showered with expensive gifts.[9]

In 1910, the Jewish Hospital had been described as the wealthiest hospital in peripheral Russia; now notices appeared in local newspapers asking for donations to keep it afloat. At the children's hospital, meanwhile, overcrowding bred tragedies of its own. 'Is the nurse guilty?' demanded a headline in the *Odesskiy listok* (*Odessa Sheet*), one of the city's main daily newspapers. A feverish child had died after falling from a second-floor balcony, and a nurse had been blamed. The author of the article felt inclined to pardon her: there were seventy-five sick children on two floors, at the hospital, and only two nurses to look after them. The nurses were working around the clock. They couldn't possibly keep track of all of the children, all of the time.

Stefansky monitored the epidemic throughout the autumn. Though most people saw out their illness at home, he judged on the basis of hospitalisations that the autumn wave peaked in the city around the end of September. On 8 October, Bardakh announced that the epidemic had passed its peak, allowing the organisers of the black wedding to claim that their efforts had paid off. He predicted that cholera would die out with the arrival of the cold weather, and that Spanish flu would last only a little longer – and he was right on both counts. When Odessans learned from their newspapers, in the second week of October, that the British prime minister David Lloyd George had been laid low with *ispanka*, someone suggested organising a black wedding for him. A local rabbi responded that there wasn't any point, since the ritual only worked *in situ*, not at a distance.

In November the armistice was signed and the Germans and Austrians left the city. Nationalist Ukrainian forces took power in Kiev, but for several weeks different factions vied for control of Odessa and Yaponchik's gangsters continued to exploit the power vacuum. Electricity was intermittent, trams no longer ran and fuel was in short supply, but the hospitals continued to function despite haemorrhaging staff. Doctors thought that *ispanka* was behind them. On 22 November, Bardakh told the Society of Odessa Doctors that it had been worse than the epidemic of the 1890s, the so-called 'Russian' flu. He added that the Spanish variety had distinguished itself by the abundance of nervous and respiratory complications that had accompanied it. In December the French arrived and, with the help of White Russian forces, cleared Odessa of Ukrainian troops. The city was now so swollen with refugees that it was like a 'packed bus', and because it was cut off from internal supply lines, food prices rocketed.[10] Soup kitchens were opened for the poor. The Zionist Pinhas Rutenberg, who passed through Odessa in early 1919, remembered it as a time of 'insanely growing expense, hunger, cold, darkness, pestilence, bribery, robbery, raids, killings'.[11]

Despite or because of an imminent sense of doom, Odessans continued to pursue their pleasure, and in the midst of all the killing and carousing, the Spanish flu returned. In early February, Vera Kholodnaya lent her star aura to a charity concert at the Club for Literature and Arts, to raise money for unemployed artists.[12] With her co-star, Ossip Runitsch, she performed an extract from their film, *Last Tango*. It was cold in the club, the audience huddled inside their furs, but she wore only a flimsy evening gown. As she was returning to her hotel after the show, the horse pulling her carriage stumbled, and she had to walk the rest of the way. The next day, she was ill. None of the illustrious doctors called to her bedside could save her, and she died eight days after her last performance. Her family requested that she be embalmed, so that her body could be transported back to her home city of Moscow, once – as they believed would happen – the old regime had been restored. The pathologist who ran the morgue at the Old City Hospital, M. M. Tizengausen, performed the honours, entering *ispanka* as the cause of death on her death certificate.

On 18 February, two days after Kholodnaya's death, the traditional Orthodox service of prayers for the departed was held at the city cathedral. The crowd was enormous, and it included a number of Jews. A struggle ensued: neither the presiding priest nor certain members of the actress's circle wanted them there. They refused to leave, however: they wanted to pay their respects to the beautiful star who had delighted them too. A more senior priest settled the matter when he ordered the service to go ahead and the Jews be allowed to stay.

Her funeral took place the next day, also in the cathedral, and it was captured on camera – as was only fitting in the temporary capital of the Russian film industry. One journalist who was present wrote later that he had felt as if he were on the set of a movie in which the queen of the screen was starring. He recalled the last film he had seen of hers, and how the audience had burst into applause at her first appearance. The cathedral was once again full, and crowds lined the route to the First Christian Cemetery, where Kholodnaya was to be placed in the crypt of the cemetery church, pending the move to Moscow. Her open coffin was carried by some of her admirers, and inside it she lay in the dress she had worn in one of her most popular films, a tragic romance called *U Kamina* (*By the Fireside*).

Kholodnaya's remains never reached Moscow, and at some point they vanished. The most likely explanation is that they were still in the church at the First Christian Cemetery when it was destroyed in the 1930s, and the site asphalted over. But the mysterious disappearance fuelled the many conspiracy theories that surrounded her death, and that still surround it to this day. According to one such theory, she was poisoned by white lilies – her favourite flower – brought by a French diplomat who suspected her of spying for the Reds. Within a few days of her funeral, the film of it was being shown in the same cinemas that had shown *The Woman Who Invented Love* the previous summer, while Paustovsky wrote that Yaponchik's bandits, sated with looting, crowded Odessa's nightclubs, 'singing the heart-rending lay of Vera Kholodnaya's death'.

The war and the plagues over, the indefatigable Bardakh turned his home into the local headquarters of a nationwide campaign to eradicate typhoid and cholera. He continued to conduct his own research, despite the ongoing shortages, adapting as always to circumstance. 'The winter of 1921–2 was harsh in Odessa, the laboratories were not heated,' he wrote. 'As a result, it was only possible to study the bacteria that could develop under very low temperatures.'[13] Under his guidance, the city's Novorossiya University became one of the Soviet Union's leading centres for bacteriology.

After his death in 1929, he was buried in Odessa's Second Jewish Cemetery, 'among the Ashkenazis, Gessens and Efrussis – the lustrous misers and philosophical *bons vivants*, the creators of wealth and Odessa anecdotes', as Babel described it. The cemetery was demolished in the 1970s, and its occupants consigned to oblivion. Only a few were spared after their families protested, and they were transferred to the Second Christian Cemetery. Among them was Bardakh, whose grave now stands next to that of another prominent Odessan Jew in a sea of Orthodox crosses. The other Jew was Mendele Mocher Sforim, the writer who, in his *Tales of Mendele the Book Peddler*, described a black wedding conducted in the belief that 'the knot tied amidst the graves of the parish dead, the contagion would at last stop'.

10

Good Samaritans

Your best chance of survival was to be utterly selfish. Assuming that you had a place you could call home, the optimal strategy was to stay there (but not immure yourself), not answer the door (especially to doctors), jealously guard your hoard of food and water, and ignore all pleas for help. Not only would this improve your own chances of staying alive, but if everyone did it, the density of susceptible individuals would soon fall below the threshold required to sustain the epidemic, and it would extinguish itself. In general, however, people did not do this. They reached out to each other, showing what psychologists refer to as 'collective resilience'.[1]

'That there was a great many robberies and wicked practices committed even in this dreadful time I do not deny,' wrote Daniel Defoe in 1722, describing the 1665 outbreak of plague in London. But he went on to describe how 'The inhabitants of the villages adjacent would, in a pity, carry them food, and set it at a distance, that they might fetch it, if they were able.'[2] A similar pattern was observed during the Spanish flu. There were certainly examples of antisocial behaviour. One Major Wells, a police officer in south-west Tanzania, for example, reported an increase in crime and cattle theft which he attributed to the pandemic, and reports are legion of profiteering when food, medicines or coffins ran short.[3] But on the whole, these were the exceptions that proved the rule.

The discovery that most people behave 'well' in a crisis may warm our hearts, but it reveals a fundamental irrationality in the way we think about epidemics. When the French pacifist and writer Romain Rolland – a man who had won the Nobel Prize in Literature in 1915 – developed the symptoms of Spanish flu while living at a hotel on Lake Geneva, the hotel staff refused to enter his room. If it hadn't been for the devoted attentions of his elderly mother, who happened to be visiting him, he might not have survived. We are tempted to condemn the hotel staff for their callous attitude, but in fact their actions probably limited the spread of the disease, and perhaps even saved lives. They unwittingly imposed a very small, very localised sanitary cordon around the unfortunate Rolland.

Doctors tell us to keep away from infected individuals during an outbreak, yet we do the opposite. Why? Fear of divine retribution might be one factor, especially in earlier eras. All three major monotheistic religions – Islam, Judaism and Christianity – insist on the importance of family, charity and respect for others. Fear of social ostracisation once the sickness has passed might be another. Or perhaps it is simple inertia: in normal times, or even in the context of a different kind of disaster – an earthquake, say – to help others might be the most appropriate response. It's only contagion that turns that rationale on its head, but we are too slow, or perhaps too dazed, to work it out. Psychologists suggest an even more intriguing explanation, however. They think that collective resilience springs from the way people perceive themselves in life-threatening situations: they no longer identify as individuals, but as members of a group – a group that is defined by being victims of the disaster. Helping others within that group, according to this theory, is still a form of selfishness, just selfishness based on a broader definition of self. It is the idea that we are all in this together. And it makes no difference if the disaster is an earthquake or a flu pandemic – only in one case the response is rational, and in the other it is not.

Take health workers, for example. These people are in the front line of any epidemic, and governments often worry that they will desert their posts as soon as they see that their own lives are in danger, and renege on their 'duty to treat'.[4] The Spanish flu showed the opposite: most doctors stayed at work until they were no longer physically able to, or until they posed a risk to their patients. 'Then the flu hit us,' wrote the poet-doctor William Carlos Williams, in Rutherford, New Jersey. 'We doctors were making up to sixty calls a day. Several of us were knocked out, one of the younger of us died, others caught the thing, and we hadn't a thing that was effective in checking that potent poison that was sweeping the world.'[5]

'We were all in the same boat, tossed about on pestilent seas, sick at heart and frustrated,' wrote Maurice Jacobs, a physician in Hull, England. 'More than one doctor expressed the intention of committing some minor crime with the object of being locked up for the duration of the epidemic. Needless to say the idea was never carried out in practice.'[6] In Japan, volunteers from the Tokyo Doctors' Medical Association gave free vaccinations at night to the poor and the *burakumin* (outcasts), while in Baden, Germany, the Catholic Church set up a programme to train young women as nurses. These women, who were required to make home visits, apparently took up their duties with enthusiasm, because in 1920 an anonymous German doctor complained about overzealous Catholic nurses exceeding their competence and being a nuisance to rural doctors.

When there were no doctors, missionaries, nuns and other religious figures took up the slack, and when they weren't available, ordinary people stepped in – even if, normally, they were divided by deep social gulfs. One of Richard Collier's correspondents, a white South African, wrote that his infant sister's life had been saved by the mother of a 'coloured' family who lived next door to them, in a rural part of the Western Cape. When both his parents fell ill, this woman – who was breastfeeding her own child – took the baby and fed her until they recovered.

Again, there were exceptions, but it's interesting to see who they were. 'Hospital sweepers deserted and refused to go near "the white men's plague", as they called it,' one British soldier wrote of his experience of recovering from the Spanish flu in India. If they had worked in a hospital for more than four years, the sweepers probably had memories of the British response to the plague outbreak that killed 8 million Indians between 1896 and 1914. They knew that they could expect no solidarity from the British. Likewise, the convicts hired to dig graves in Rio de Janeiro, and who – if the rumours were true – committed all manner of heinous crimes in the presence of the corpses, probably felt they had nothing to lose.

At some point, according to the theory of collective resilience, the group identity splinters, and people revert to identifying as individuals. It may be at this point – once the worst is over, and life is returning to normal – that truly 'bad' behaviour is most likely to emerge. The Swiss Red Cross, which had been gratified by the surge of unqualified women volunteering to nurse, lamented the fact that some appeared to have done so for 'morally dubious' reasons. These impostors often clung to their new roles even after the epidemic had passed, it reported, 'presenting themselves as experienced nurses, donning diverse uniforms and sometimes producing fake medical certificates designed to trick the public and the medical corps'.[7]

In 1919, Carnival in Rio took the theme of divine punishment, and more people attended it than ever before. Flu had not quite vanished from the city, and death was still very present. Carnival songs fixed the trauma for perpetuity, and some of the *blocos* or neighbourhood carnival groups gave themselves flu-themed names – 'The Block of the Holy House', 'The Block of the Midnight Tea'. A change came over the revellers on Carnival Saturday – a desire for catharsis, perhaps. The newspapers documented the 'unusual joy' that engulfed the city. 'We had a party,' wrote one chronicler, with droll understatement; 'the binge was full', another. 'Carnival began and overnight, customs and modesty became old, obsolete, spectral . . . Folk started to do things, think things, feel unheard-of and even demonic things.'[8]

Something similar may have happened in the wake of the Black Death in the fourteenth century. 'Nor is it the laity alone who do thus,' wrote Giovanni Boccaccio in *The Decameron*, describing that interlude in Florence. 'Nay, even those who are shut in the monasteries, persuading themselves that what befitteth and is lawful to others alike sortable and unforbidden unto them, have broken the laws of obedience and giving themselves to carnal delights, thinking thus to escape, are grown lewd and dissolute.'

In Rio, in that unusual atmosphere, boundaries became blurred. There are references to numerous *defloramentos* – deflowerings – which led in turn to a cohort of children dubbed 'sons of the flu'. Such reports are hard to confirm, but one historian, Sueann Caulfield, has scoured the archives and found that, in the period immediately after the epidemic, there was indeed a surge in reported rapes in Rio, to the extent that they temporarily outnumbered other types of crime.[9] Some saw this wave of obscenity as the revenge of the unloved dead; others as a shocking reassertion of an inextinguishable life force. Whatever it was, it brought closure: the pandemic was over. Humanity had entered a post-flu world.

WOLVES ON THE PROWL

Perhaps the best illustration of both the 'best' and the 'worst' of human behaviour is to be found in Bristol Bay, Alaska. When the Spanish flu swept through Alaska in the autumn of 1918, two groups of Eskimos were spared: those living in the outer islands of the Aleutian chain, the furthest west you can go in North America without getting your feet wet, and the Yupik of Bristol Bay. The Aleuts had a natural cordon sanitaire in the Pacific Ocean, but Bristol Bay, the eastern-most arm of the Bering Sea, was remote in a different way. Bound by the Alaska Peninsula to the south, several mountain ranges to the north and a waterlogged tundra in the interior, it isn't easily accessible today, and was even less so when steamer and dog team were the only means of transport. In winter the Bering Sea has a tendency to freeze over, blocking the ocean route entirely. But when, in the spring of 1919, the sea ice began to break up and the first fishing boats of the season arrived, the flu came with them.

'The surroundings are truly arctic,' wrote Katherine Miller, a Seattle-trained nurse who laid eyes on Bristol Bay for the first time that spring. 'No vegetation except the grasses and mosses of the vast swampy plain or tundra which extends limitless on every side.'[10] A priest who had explored the Alaskan coast two winters earlier was only a little more generous: 'In the main the country traversed is as dreary and naked as I suppose can be found on earth, and cursed with as bitter a climate; yet it is not without scenes of great beauty and even sublimity, and its winter aspects have often an almost indescribable charm; a radiance of light, a delicate lustre of azure and pink, that turn jagged ice and wind-swept snow into marble and alabaster and crystal.'[11]

In fact Bristol Bay is subarctic, not arctic. The summers can be warm, if short, but in winter the temperature can drop below minus forty degrees Celsius. The country may have struck southerners as inhospitable, but it was rich in natural resources. The rivers that empty into Bristol Bay are the greatest spawning grounds for red salmon in the world, as Captain Cook intuited when he passed by in 1778, on his fruitless quest for the Northwest Passage. Eyeing the mouth of a river, he imagined that 'It must abound with salmon, as we saw many leaping in the sea before the entrance; and some were found in the maws of the cod which we had caught.'[12] The land, meanwhile, was home to bear, moose and caribou. The Yupik were less nomadic than other Alaskans – they had much of what they needed on their doorstep – and for that reason, combined with their isolation, contact with outsiders came relatively late for them.

For millennia their lives had followed the seasons. From the first snowfall in October they gathered in their villages, to see out the winter living off stocks they had laid in during the warmer months. In the spring they dispersed in small family groups to hunt or trap game, building themselves temporary brush or canvas shelters to live in, and by June they were back in their villages to fish salmon. The men left again in August, to hunt until the snow came.

Their villages consisted of *barabaras*, dwellings made of turf covering a log frame, two-thirds of which were underground. The women and children lived in smaller *barabaras* circling a large, central one known as the *qasgiq*. The *qasgiq* was a male domain, the place where the single men slept, but in winter it often became a communal space – a space where, as anthropologist Margaret Lantis wrote in 1950, the dark days and nights were spent 'delighting the spirits of the animals with feasts, dances, and masks'.[13] The Yupik inhabited a world crowded with spirits, both human and animal. As one elder explained, 'When the Yupik walked out into the tundra or launched their kayaks into the river or the Bering Sea, they entered into the spiritual realm.'[14]

The first to intrude on this world were the Russians. In 1818 they established a fur-trading post at Alexandrovsky Redoubt at the mouth of the Nushagak River, which spills into an arm of Bristol Bay – the site of the modern town of Dillingham. In 1867, America bought Alaska from Russia and within a few decades the commercial fishing industry had taken off in the bay, under the auspices of the San Francisco-based Alaska Packers' Association (APA). The Russians brought the Orthodox religion, the Americans the Protestant one, and both brought disease – a series of devastating epidemics that culminated, in 1900, in the most lethal of them all: a double epidemic of flu and measles, known to Alaskans as the Great Sickness, that wiped out between a quarter and a half of the Eskimos of Western Alaska.

By 1919, the Yupik were a people in transition. They still lived mainly by hunting and fishing, and sought out shamans to interpret the spirit world for them – especially when they were sick – but many now lived in modern houses, wore store-bought clothes and, in the Nushagak area, professed the Orthodox faith. In the summer of 1918, the salmon run failed – due to overfishing, in the opinion of the local Bureau of Fisheries warden – meaning that Yupik stocks were low the following winter, and they were hungrier than usual, come the spring.

The flu entered Alaska at Unalaska Island, one of the most landward islands in the Aleutian chain that forms the tail of the Alaska Peninsula, and hence a natural stopping point for north-bound ships. The story of how it spread from there, north-eastward to Bristol Bay, is the stuff of Alaskan legend. A Russian priest, Father Dimitri Hotovitzky – known to his flock as Father Hot Whiskey – travelled from Unalaska up to the bay to lead celebrations of Orthodox Easter, and it is said that whoever attended his services returned home sick.[15] That he infected the bay's population is possible, but unlikely. The incubation period for flu, during which a person can be infected but symptom-free, is between one and four days. Orthodox Easter fell on 20 April in 1919, coinciding as it sometimes does with 'western' Easter. The first cases in Bristol Bay were reported around 12 May, three weeks later. Even allowing that some earlier cases may have gone unreported, three weeks is an unreasonably long incubation period. Someone who followed in Hot Whiskey's footsteps more likely brought the virus in.

Alaska was a US territory in 1919, not yet a full state. The territorial governor, Thomas Riggs, therefore had no vote in Congress, and his voice was drowned out by the louder ones of the representatives of the then forty-eight states. Riggs had managed to persuade the government to provide funds for a territory-wide quarantine during the autumn wave of 1918, but it had been lifted in March, and when the disease reappeared a few months later, his renewed pleas fell on deaf ears. In the forty-eight states, this third wave was relatively mild. The burden of managing the new epidemic in Alaska therefore fell on the doctors employed by the APA at salmon canneries around the bay, and on the government hospital at Dillingham.

This hospital was run by a doctor named Linus Hiram French. He knew and loved that part of Alaska, having worked there previously as a cannery doctor. After taking up his government post in 1911, he set off to survey his vast catchment area, travelling through the winter months by dog and reindeer team, or on foot with snowshoes. On his return in the summer of 1912, he reported to his government bosses that the houses he had visited were, in the main, warm, damp and dark, 'as the native keeps in all warm air, to avoid chopping wood', and that dogs and humans shared living space. TB and syphilis were common, as was the eye disease trachoma. He treated some of the sick, sent others to the hospital, and issued instructions on how to prevent the diseases that were preventable. He was surprised to find that many of the people he encountered thought that Alaska was still Russian: 'In every house are hung pictures of Russian priests or the tsar and all keep time by the Russian calendar.'[16]

As soon as the flu appeared French imposed a quarantine on the region. Those Yupik who had not yet reached their villages for the start of the fishing season therefore found themselves cut off from them, and if they had passed through infected areas, placed in 'detention cabins' for ten days at their own expense. The APA doctors also declared quarantine zones around individual villages, and supplied afflicted ones with food, fuel and medicine. Despite these measures, the hospital at Dillingham was soon operating at full capacity, as were the makeshift hospitals the APA doctors had set up by erecting tents over wooden platforms. In late May, as the epidemic peaked, both French and the two nurses assisting him fell ill, and French wired the captain of a US Coast Guard cutter, the *Unalga*, requesting urgent assistance.

The *Unalga* had left San Francisco over a month earlier on one of its routine cruises to patrol the coastline, and incidentally, to ferry passengers, mail and goods between stopping points along its route. The ship's captain, Frederick Dodge, knew Alaska well, but it was the first cruise in those waters for the *Unalga*'s new cook and watch officer, the unfortunately named Eugene Coffin. He later noted in his diary that Captain Dodge had a penchant for the Russian icons and samovars that were to be found in many Alaskan homes, and that he collected along the way: 'I guess he paid something for the things, no doubt.'[17]

The *Unalga* had a doctor on board, and when it reached Unalaska – the main town on Unalaska Island – on 26 May, the crew found the town in the grip of flu. Captain Dodge organised a relief operation, and in his diary entry for 30 May, Coffin wrote that, '*Unalga* feeding and nursing the entire town and burying the dead.' By then, according to the ship's official log, the captain had sent a message to French saying that the *Unalga* had its hands full and couldn't come to his aid. French seems never to have received that message, and two cannery superintendents who also sent SOSs to the *Unalga* claimed never to have received replies either. By 7 June, the epidemic had passed its peak in Unalaska, but Dodge had received word from Governor Riggs that a relief ship, the USS *Marblehead*, was expected there on 16 June, carrying fresh supplies provided by the American Red Cross. He waited for the other ship to arrive.

The *Marblehead* and one other ship were the federal government's only concession to the new tragedy unfolding in Alaska, and the *Marblehead* had one passenger of note on board: Valentine McGillycuddy. McGillycuddy was a physician who had made a name for himself as an Indian agent, but an unusual one, whose sympathies lay at least partly with the Sioux he was sent to 'civilise'. He counted Crazy Horse among his friends, and was at the great Sioux chief's bedside when he died in 1877. When America joined the war he had scented new adventure, and asked the War Office to send him to Europe as a surgeon or reconnaissance officer. They declined on the grounds that he was too old. He offered his services to the Red Cross and received the same answer. Only the US Public Health Service was interested, once the Spanish flu had broken out. He was summoned to see one of its representatives in San Francisco, to whom he confessed that 'he didn't know a damn thing about influenza'. 'I can't advise you,' the representative replied. 'Not one of us knows a damn thing about it, either.'[18] Thus the seventy-year-old doctor came out of retirement, first to fight the flu at the New Idria mercury mines in California – where he observed the supposedly prophylactic effects of mercury vapour – and now in Alaska.

The day after the *Marblehead* arrived at Unalaska, McGillycuddy – along with two other doctors, three pharmacist's mates and four nurses – boarded the *Unalga*, taking some of their supplies with them, and the cutter set off on the two-day journey to Bristol Bay. 'As the ship reached port, the doctor stood on the deck of the launch and scanned the coast,' wrote Julia Blanchard McGillycuddy, the doctor's wife and biographer. 'A gentle breeze blew offshore bringing with it a cadaveric odour. There was something wrong, the doctor said, not far inland.'[19]

The *Unalga* anchored off Dillingham on 19 June. The *Marblehead*, which had followed it, headed for a different part of the bay with the remainder of its medics and supplies on board. Both were 'too late to be of any service', according to one cannery doctor, because by then the worst was over. French and the two nurses at the government hospital, Rhoda Ray and Mayme Connelly, were back on their feet, and two more nurses had arrived from the Alaskan port city of Valdez. They had completed the 800-kilometre journey by boat and dog sled, and one of them, the aforementioned Katherine Miller, recorded her observations on reaching Dillingham: 'Here and up the Wood River [another tributary of Bristol Bay], the ravages of influenza were most severe. Some villages were completely wiped out . . . Whole families were found by relief parties lying stricken on the floors of their huts.'[20]

The *Unalga*'s log recorded that the crew administered aid where it was needed, but the local fisheries warden gave a different version of events. He reported that the cutter would anchor off a stricken village and send a landing party ashore which, rather than administer succour, would go hunting for souvenirs: 'Eskimo houses were invaded – in some instances rifled – and acts bordering on vandalism committed.' At Dillingham, the warden wrote, the four nurses from the *Unalga* did report for duty. 'They were not there an hour, however, before they invited the two nurses at the government hospital to a dance on board the cutter that evening.'[21]

Ray and Connelly explained to the four nurses that even with the extra help from Valdez, they were stretched to the limit, looking after the sick and the growing influx of orphaned children, while also taking care of the laundry and keeping the hospital clean. They declined the invitation and the visitors left. When they returned a couple of days later, Ray and Connelly told them their services were not required, since they didn't need any more mouths to feed. The fisheries warden was complimentary about one unnamed doctor in the relief party, probably McGillycuddy, who took temporary charge of the hospital, showing 'efficiency and devotion to duty', and freeing French to go out to the villages.

The *Unalga* had not covered itself with glory in Bristol Bay, but it did have one final contribution to make. On 25 June, a party that included Coffin and McGillycuddy headed up the Wood River on French's launch, the *Attu*. In the early hours of the following day they came in sight of a village, probably Igyararmuit, meaning 'people who live at the throat', because it was situated close to where the river flows out of Wood Lake. The *Attu* tied up to a government barge that was there for the purposes of a salmon census, and those aboard tried to snatch some sleep despite being strafed by mosquitoes. In the morning they went ashore and found the village deserted. A bad smell was coming from one of the *barabaras*, and they ventured in to investigate. Coffin described what happened next: 'On going into the low narrow door into the first of two connecting rooms was unexpectedly confronted by three big malamutes – promptly retired closing the door – broke windows on the roof and shot the dogs – two skulls and many large bones all picked clean scattered over floor and evidence that the dogs had been fighting over the remains.'[22] It was a ghoulish echo of another American's observation, during the Great Sickness of 1900, that, 'Prowling dogs ate at dead bodies while from the foothills came the long drawn-out eerie calls telling that wolves were near.'[23]

The party returned later in the day to sprinkle the village with kerosene and set fire to it, shooting three more dogs the size of timber wolves. Once the fire had taken, they headed back downriver, and on 28 June the *Unalga* set a course for Unalaska. 'All hands glad of it,' wrote the cook, who would return twice more to the Bering Sea, though never again with Captain Dodge. Three days later, the *Marblehead* steamed south for San Francisco, bringing McGillycuddy's Alaskan adventure to an end. For the next twenty years of his life, until his death at the age of ninety, he worked as the house doctor at the Claremont Hotel in Berkeley, California.

The epidemic tailed off over July, by which time it was clear that the salmon run had failed again. Bristol Bay, the region of Alaska that was affected worst by the Spanish flu, had lost around 40 per cent of its population, and the Yupik who survived would recall that period as 'Tuqunarpak', which translates roughly as 'big deadly era'. The Nushagak area seems to have been particularly hard hit. Some villages, Igyararmuit among them, had simply ceased to exist; others were so devastated that their remaining inhabitants abandoned them. During his 1912 expedition, French had counted nineteen villages along the Nushagak River, that varied in size from fifteen to 150 inhabitants (only three of them were marked on the map). Assuming an average of seventy inhabitants, that gives an estimated total population of 1,400. In 1920, Father Hotovitzky reported that, in Nushagak parish, 'No more than 200 parishioners were left in all the villages.'[24]

Apparently oblivious to the ugly rumours that had dogged him on his pascal peregrinations, Hotovitzky had prepared an audit of the Aleutian Deanery for His Eminence Alexander Nemolovsky, Archbishop of the Aleutian Islands and North America. Despite the benevolent intervention of martyr and healer St Pantaleon, he explained, the parishes for which he was responsible had been much reduced in 1919. 'Those parishioners who survived saw out the year piously, by the Grace of God,' he wrote, adding: 'In Nushagak itself, the church was closed because there are no Orthodox left. During the epidemic, many objects were stolen from the church by the Americans.'[25]

Close to 150 orphans were rescued from all points of the bay while the epidemic was raging. 'They were freezing and shivering in cold huts, without fires or food, with little clothing or bedcovering – many of them crying, huddled about their dead,' reported one APA superintendent.[26] More were discovered after the epidemic had receded, and though the figures are unreliable, the eventual number of orphans that was brought to the hospital at Dillingham – a town of fewer than 200 inhabitants at the time – may have been closer to 300.[27]

To begin with, the nurses' principal dilemma was how to dress them. 'Many had only clothing made from old flour sacks obtained from trading posts scattered through the vicinity,' wrote Miller.[28] French appealed for government funds to build an orphanage, and these were granted. It would be the doctor's final gesture: in the months following the epidemic, he left Bristol Bay, never to return. Nearly half a century later, an anthropologist named James VanStone who made a study of the Yupik noted that, once grown, most of the flu orphans had tended to stay in and around Dillingham, rather than return to their places of origin. Today, all the indigenous residents of Dillingham claim to be descended from them.

PART FIVE: Post Mortem

British soldiers bathing in the sea at Étaples, 1917

11

The hunt for patient zero

'We desire to present in this preliminary note a consideration of the similarity of the present epidemic to the epidemic of pneumonic plague which broke out in Harbin, China, in October, 1910, and spread rapidly and continuously throughout Northern China at that time; and to suggest that this epidemic may be the same disease modified by racial and topographic differences.'[1]

So wrote James Joseph King, a captain in the US Army Medical Corps, on 12 October 1918. Even in 1918, medics doubted that Camp Funston – the military base in Kansas where cook Albert Gitchell fell ill on 4 March 1918 – was the origin of the 'Spanish' flu. Alternative theories emerged while the pandemic was still raging, and initially they pointed to China. Where Captain King led, others followed. The quick pointing of fingers to the east was probably influenced – albeit often unconsciously – by contemporary western attitudes towards the peoples of East Asia, known collectively as the myth of the 'Yellow Peril'. At its most extreme, this xenophobia manifested itself in accusations that Asians were to blame for falling birth rates in Europe, rising criminality, the kidnapping of women for the white slave trade and even vampires (who were supposed to have reached Transylvania from China via the Silk Road).[2]

Captain King was undoubtedly sincere, but the possibility didn't even occur to him that the pandemic might have been seeded in his own country. Americans were, naturally, only victims. 'Since our soldiers and sailors have been returning from the battlefields of France,' he wrote, '[the disease] has become very prevalent and serious in our camps and cities all over this country.' However, the Chinese-origins theory has been revived in recent years, in the light of new historical evidence regarding the role that China played in the war. The Yellow Peril notwithstanding, it remains a possibility that the pandemic did start in the east, and to understand why we have to go back to an outbreak of disease in Manchuria in 1910 – the very outbreak, indeed, that King referred to in his 'preliminary note'.

China in 1910 was known as the sick man of Asia. It was sick in the real sense, having a gargantuan public health problem, and it was sick in the metaphorical sense, having bled land and autonomy to foreign powers since the middle of the previous century. The outbreak in the sensitive frontier region of Manchuria dissolved the tenuous distinction between the real and the metaphorical, and when news of it reached the mandarins in Peking, they recognised it for what it was – the first, distant death knell for the ruling Qing dynasty. Revolution was in the air, and the empire was weak. Russia and Japan had already run rail lines into mineral-rich Manchuria, and Japan had recently annexed Korea, so that it now shared a border on the mainland with its ancient foe. A plague that posed a threat not only to those nations, but also to Europe and America, which had their own interests in China, would provide them with the pretext to invade – with men in white coats leading the charge. The mandarins knew they had to bring the plague under control without foreign intervention. They had to put it in the hands of a doctor they could trust – one of their own. The man they chose was Wu Lien-teh (Wu Liande).

The son of a Chinese goldsmith, Wu was born in Penang, a British colony in what is now Malaysia, in 1879 and graduated from Cambridge University in 1902 – the first medical student of Chinese descent to do so. He went on to study with Mechnikov in Paris, and Koch's student Carl Fränkel in Halle, Germany. In 1908, following his return back east, he took up a post training military doctors at the Imperial Army Medical College in Tientsin. This is where he was when, in November 1910, he received a telegram from the Ministry of Foreign Affairs, ordering him to go north and rein in the epidemic.

When Wu arrived in the Manchurian city of Harbin, near the Russian border, he found conditions there unsatisfactory. 'The local magistrate was a confirmed opium smoker, prided himself upon being an amateur physician and did not believe in germs or foreign medicines,' he later recalled.[3] There were no hospitals, only 'filthy' plague houses into which suspected cases were thrown. Many people had already fled in panic, and others were preparing to travel south to celebrate the Lunar New Year with their families. Wu suspended all non-essential train travel and turned schools, theatres and bathhouses into disinfection stations. Temples and deserted inns became plague hospitals, idle train wagons isolation wards. Seven hundred police and a thousand soldiers were put at his disposal, and he used them to enforce house-to-house searches and quarantine. The Manchus were less than cooperative. Terrified of quarantine, and justifiably – they had seen there was little chance of returning from it – they were also bound by the obligations of filial piety. They would often fail to report a case while the patient was alive, and sometimes try to hide the corpse when he or she was dead.

Wu quickly suspected that he was dealing with pneumonic plague. His patients were reporting fever and chest pain. Soon they were coughing up blood and their skin had taken on a purplish hue. Nobody who fell sick survived, and death typically came within a few days. Merely suspecting plague was not enough, though. He knew that to identify the disease definitively he would have to isolate the plague bacterium, which meant performing an autopsy. In pre-revolutionary China, violating a corpse was a serious crime, one that could itself be punished by death, so it's an indication of the seriousness the mandarins accorded the epidemic that they granted him an imperial dispensation to do just that. Having performed the autopsy on the body of a female Japanese innkeeper near Harbin, he analysed the bacteria cultivated from her lung tissue and found that she had indeed been infected with *Yersinia pestis*. Meanwhile, the bodies of plague victims were piling up outside the city. The temperature was minus twenty, the ground frozen solid; burial was out of the question. Wu obtained another dispensation to cremate the corpses – another practice wholly contrary to Chinese custom – and the pyres burned for two days throughout the Lunar New Year at the end of January.

The epidemic petered out in April, and Wu's imperial masters were delighted. Though it had spread as far south as Hopei (Hebei) and neighbouring Shantung (Shandong) provinces, though it had claimed 60,000 lives, it had not breached Chinese borders. The threat of invasion had been averted. 'The high rank of major of the Imperial Army with blue button was conferred upon me overnight,' Wu boasted, 'so as to enable me to receive imperial audience without unnecessary formalities.'[4] The reprieve for the Qing was short-lived, however. The following October, the dynasty was overthrown and a Chinese republic was born. The diminutive, silver-tongued Wu (he stood five feet two, or one metre sixty, in his stockinged feet) found favour with the new regime, and in December 1917, he was called out to deal with another deadly epidemic of respiratory disease.

This time, the outbreak had been reported in Shansi – Governor Yen's fiefdom – and among Wu's fellow plague-fighters was the missionary Percy Watson. Wu was about to discover that his ideas were no more popular in the countryside than they had been seven years earlier, least of all in conservative Shansi. When he tried to perform an autopsy without first asking permission from relatives of the deceased, an angry crowd surrounded the coach that served him for accommodation and set fire to it. It was this incident that persuaded Watson not to perform an autopsy in Wangchiaping one year later – the autopsy that might have enabled him to make a definitive diagnosis – 'because of the great trouble caused in Northern [Shansi] last year when Dr Wu Lien-teh got such a specimen'.

Wu escaped and fled to Peking, taking with him the couple of tissue samples he had managed to obtain, and on 12 January 1918 he announced that he had found the plague bacterium in them. Other doctors who had been to the epicentre of the outbreak immediately contested his diagnosis, as did officials in Shansi. Though it certainly bore many of the hallmarks of that disease – bloody sputum, chest pains, fever – they deemed it milder than the epidemic of 1910. Strikingly, death was the exception rather than the rule. The officials insisted that it was merely a severe form of 'winter sickness' – something more like influenza.

If it was influenza, one thing is certain: Wu would have had no way of demonstrating it. Still, he claimed to have seen the plague bacterium. Some have suggested that he exaggerated his confidence in his diagnosis to convince the authorities to put in place the containment measures he considered so vital, or more simply, because he had already convinced himself that he was dealing with plague. Whatever the truth, a doubt hangs over the nature of the disease that ravaged Shansi in the winter of 1917, and that doubt has fuelled speculation that it was in fact the first manifestation of the Spanish flu. If so, how did it travel from isolated Shansi to the rest of the world? According to the revived theory of an eastern origin, the Chinese Labour Corps (CLC) provides the key.[5]

While the epidemic raged in Shansi, war raged on the other side of the world. China had declared itself neutral in 1914, its hands being tied by the fact that warring nations on both sides claimed concessions within its borders (it finally declared war on Germany in August 1917). Right from the start of the war, however, its leaders had tried to find a way of contributing without compromising that neutrality, in order to earn a place at the negotiating table when the inevitable peace process happened. They saw that process as their chance to claw back the territory the last Qing emperors had ceded to foreign powers. The plan they came up with, in cooperation with the British and French governments, was to create a body of labourers who would not take part in combat, but who would take on the heavy lifting behind the lines – digging trenches, mending tanks and assembling shells.

This was the CLC, and beginning in 1916, in a largely secret operation, as many as 135,000 men were transported to France and Belgium under its auspices, while another 200,000 went to Russia.

These men were carefully selected from northern Chinese populations who were considered taller, on average, and more suited to a cold climate than southerners. Most were peasant farmers from Shantung and Hopei provinces, though some came from as far away as Shansi. Hopei is sandwiched between Shansi and coastal Shantung, and all three provinces were affected by the 'plague' of the winter of 1917. The British often used missionaries to recruit them. American journalist and secret agent Josef Washington Hall was travelling in Shantung when he came across one 'recruiting coolies by means of his marvellous oratory in the temple square'. The priest, famous in those parts, was known by his Chinese name, Pastor Fei. Hall recounted that Fei Mu-sa told the crowd:

> I have come to tell you of an opportunity to see the world. Those of you who are able-bodied shall sail across two seas to the land where men look the opposite way from you to see the sky, where there are buildings as large as a walled village, in cities as clean as a threshing floor. You shall work there only one-third of each twenty-four hours, and each receive the pay of three men, while your families will be paid their food money each month here at home. You will be safe from danger, for iron masters as large as three-beam houses will protect you. And when the great British king has won victory he will send you back to your homes with enough money to buy you each a new field, and a reputation which will make you esteemed of your neighbours and posterity. All this I swear by my honour. If it is not true, when you come back, look me up.[6]

It wasn't true, unfortunately – though history doesn't relate if those whom the pastor swayed did look him up on their return.

They would be mistreated as racially inferior 'chinks', exploited, and not always kept at a safe distance from the front line. From the spring of 1917 they were recruited mainly at Tsingtao (Qingdao) in British-occupied Shantung, where they were subjected to a medical inspection before being dispatched around the world. This inspection was quite rigorous, until the numbers of recruits grew very large and the system started to break down, but it was designed to weed out mainly diseases considered 'Asian' – such as trachoma, that can cause blindness – not common-or-garden flu (for which, anyway, they had no test). Those labourers destined for France or Belgium went eastward via Canada or westward via the Cape of Good Hope. If they took the easterly route, they entered Canada at Victoria, British Columbia. The voyage took three weeks, they were packed like sardines into poorly ventilated holds, and conditions at the William Head Station on Vancouver Island where they were quarantined weren't much better. Herded onto sealed trains that were protected by armed guards, they were then ferried across the country to Montreal or Halifax, from where they embarked on the final sea voyage to reach the killing fields of Europe. Those who went west entered France at Marseilles.

Slivers of circumstantial evidence exist to support the Chinese-origin theory. The numbers of men at the Tsingtao depot swelled over the winter of 1917–18, and by January many of them were complaining of sore throats. Something like flu was in the air when Pastor Fei was recruiting in Shantung. Although Hall doesn't mention the exact date when he saw him, it was in the spring of 1918, and that night Hall woke with a chill. 'The next morning,' he wrote, 'I had every symptom of the influenza or "small plague" as the Chinese call it, though it has killed a million or two of them.' Thousands of CLC recruits left Tsingtao that spring, and there is some evidence of a spike of respiratory disease in the soldiers assigned to guard them on Vancouver Island. This might have been nothing more than seasonal flu, but either way, the soldiers mixed with the local civilian population and could have passed it on.

There is nothing more than circumstantial evidence, however, because we don't know what the disease was that erupted in Shansi in late 1917, and receded the following April, having claimed an estimated 16,000 lives. Wu Lien-teh came the closest to identifying it, but fairly or unfairly, a shadow hovers over his credibility, and since the tissue samples he risked his life to obtain no longer exist, as far as we know, it will hover there forever.

The Chinese theory stood alone for a long time, but then in this century, two rival theories were proposed. According to one of them, patient zero or the index case – the first person to contract the 'Spanish' flu – did not fall ill in China, or even in the silent spaces of the Eurasian steppe, but a short train ride from the Western Front, in the heart of the European theatre of war.[7]

Between 1916 and the end of the war, Britain delivered a million or more fighting men to the Western Front – the sixteen-kilometre-wide system of trenches that gashed France from the Belgian to the Swiss border – but this feat presented them with certain logistical challenges. While the French, Germans and Russians had thousands of square kilometres in which to billet their reinforcements, stock their supplies and tend their sick and wounded, the British had to squeeze their entire support operation into the narrow strip of land between the front and the Atlantic Ocean. The solution they came up with was to build a camp at Étaples, a small fishing port just south of Boulogne-sur-Mer.

You can still see vestiges of the camp at Étaples. It starts at the northern edge of the town and runs up the coast – tens of square kilometres of land in which the remains of ammunition dumps occasionally break the surface. If you had flown over it in a military plane in 1916, you would have looked down on the River Canche, which flows into the English Channel at Étaples, and perhaps spied recruits being drilled in the wide dunes around it, or little huddles of deserters hiding out there. Heading north, you would have passed over the 'Bull Ring' – the notorious exercise ground where men were pushed so hard they mutinied in

1917 – shooting ranges, detention camps, and above all, row upon monotonous row of wooden barracks. Finally you would have come to the camp's northern perimeter, and been impressed, or depressed, by the sight of a dozen hospitals lined up along it. Between them these boasted 23,000 beds, making Étaples one of the largest hospital complexes in the world at the time.

On any one day, this sprawling, makeshift city accommodated 100,000 men and women. Reinforcements arrived daily from the four corners of the British Empire, and nearby were camps for German POWs and French troops from Indochina. Fifty kilometres to the south, at Noyelles-sur-Mer, near the Somme estuary, the CLC had its headquarters and a hospital of its own (Number Three Native Labour General Hospital, to give it its correct name). In all, around 2 million human beings were camped out in this small corner of northern France. By 1916, Étaples had become an overcrowded holding pen for men who knew they were about to die. The British poet Wilfred Owen, who passed through it, described the 'strange look' peculiar to the camp, in a letter to his mother: 'It was not despair, or terror, it was more terrible than terror, for it was a blindfold look, and without expression, like a dead rabbit's.'[8]

Between July and November of 1916, during the Battle of the Somme, up to ten ambulance trains a night arrived at Étaples. Many of the wounded had been exposed to mustard gas, which causes the lungs to blister. In December – a whole year before Shansi's bout of winter sickness, that is – something very like flu broke out at the camp. By the time the weather turned cold at the end of January, it had reached the proportions of a small epidemic, and it receded with the frost in March. A trio of British Army doctors led by Lieutenant J. A. B. Hammond described it in the *Lancet* medical journal in July 1917. They called it 'purulent bronchitis' and noted that it was characterised by a dusky blue hue to the face. They performed autopsies on some of the victims and found their lungs to be congested and inflamed – a signature, too, of the Spanish flu.[9]

Was purulent bronchitis a precursor of the Spanish flu? A British virologist called John Oxford thinks it was, and thanks to the assiduous record-keeping of military doctors during the First World War, he has built a persuasive case. A historian with whom he has worked, Douglas Gill, has studied the death records for British military hospitals in the French city of Rouen – a centre of hospitalisation that was almost as important as Étaples – and found that an epidemic passed through there too, at around the same time. An almost identical disease broke out in barracks at Aldershot, England, in early 1917.[10]

There is a problem with the Étaples theory, though: there are no records of outbreaks in the civilian population of northern France at that time. It seems odd that a dangerous infectious disease would erupt simultaneously at a number of military bases, while the civilian communities between them remained unaffected – especially since we know that the camp at Étaples lived 'in osmosis' with the town.[11] British soldiers 'fraternised' with local women, and frequented the town's shops, bars and brothels (the lady whose favours were most sought after called herself 'the Countess'). But there may be a simple explanation for that: under the French civilian system in operation at the time, to protect individuals' privacy, the cause of death was recorded separately from the announcement of that death. Though the public death registers have survived, often the doctors' certificates mentioning the cause of death have not. There may have been outbreaks among civilians, in other words, but if there were, no record of them survives.[12]

Hammond produced a detailed description of purulent bronchitis, but he was no better equipped than Wu to isolate a virus, so the Étaples theory, too, remains conjecture. Having proposed such an early herald event, the onus is also on Oxford to explain why the pandemic proper took so long to erupt. His suggestion is that, although conditions in northern France were highly conducive to the emergence of a new pandemic flu strain in 1916, paradoxically, they also contained it. Travel was limited to the round trip from base to front and – if you were lucky – back again, or at most to a short hop across the Channel. In the year or more's interval between the outbreak described by Hammond and the first recognised wave of the pandemic – in spring 1918 – the virus may have maintained itself in small, localised epidemics while it acquired the molecular changes that would render it highly transmissible between humans.

What if the 1918 pandemic started not in China, or in France, but further west again – just down the road from the first recorded case? The third theory suggests that patient zero was not a gassed soldier recuperating at Étaples, nor a peasant farmer labouring among the cliffs and ravines of Shansi, but a peasant farmer labouring close to the geographical heart of America – in the 'Sunflower State' of Kansas.

Camp Funston drew recruits from a catchment area that included Haskell County, 500 kilometres to the east. Haskell was one of the poorest counties in Kansas at that time. Its inhabitants lived in sod houses, grew corn and raised poultry and hogs. In January 1918, they began to fall sick, and some went on to develop pneumonia and die. A local doctor, Loring Miner, was so alarmed by the severity of the outbreak that he reported it to the US Public Health Service, even though flu wasn't a reportable disease in the US at that time. The epidemic receded in mid-March, and nobody might have given it another thought – besides the grieving inhabitants of Haskell County – except that by then, the infirmary at Camp Funston had been overrun by sick soldiers.

On the same day that the camp's chief medical officer sent a telegram to the authorities in Washington DC about *his* outbreak, 30 March, a report of the earlier one in Haskell appeared in the public health service's weekly journal. Almost nine decades passed before an American journalist, John Barry, suggested that the two might have been linked – that a young man hailing from Haskell, probably a God-fearing boy who had grown up on a farm and known no other life, unwittingly carried the virus into the midst of the American war machine, whence it was exported to the rest of the world.[13]

When you try to chart the progress of the pandemic's spring wave from the first case in Camp Funston, eastwards to France, it seems at first pleasingly linear and one-directional. Then, however, you remember that large numbers of CLC labourers were being moved across North America that spring, in specially guarded trains. Though we have no reason to believe that they had any contact with the populations they passed through, it isn't out of the question that a guard's attention lapsed momentarily, or that he took pity on a poor passenger and allowed him out to stretch his legs. His instructions were to keep the labourers moving eastwards as discretely as possible; he did not realise he was also defending a sanitary cordon. By April 1918, China was in the grip of yet another flu-like disease – an apparently new epidemic that nevertheless overlapped in time with the one that had started in Shansi the previous winter.[14] This new epidemic, according to the consensus of the Chinese medical community, was definitely winter sickness and not plague. It wasn't fatal and generally passed in four days (Wu disagreed; he was convinced it was the same disease that had broken out in Shansi, and that both were plague, but he was in a very small minority). The possibility exists, therefore, that it was the CLC that brought the flu to the eastern seaboard of North America. To confuse matters still further, there is evidence that New Yorkers were falling sick from late February 1918, *before* Gitchell checked himself into the infirmary at Camp Funston, prompting some to suggest that New York City received the infection from troops returning from France.

For the time being, therefore, all three theories of the origin of the 'Spanish' flu remain on the table. To choose between them would require a comparison of the flu strains that caused the putative precursor events, with the strain that was circulating in the autumn of 1918 – something that hasn't yet been possible. In the twenty-first century, scientists have produced a new kind of evidence that points to one of the theories being more likely than the other two – we'll come to that – but this evidence, though tantalising, is not definitive. In 2017, therefore, there is only one thing we can say with something close to certainty: the Spanish flu did not start in Spain.

Note, for now, that if the Chinese-origin theory is correct, the pandemic cannot strictly be described as a product of the war. Patient zero was a poor farmer living in a remote village in the Chinese interior, who at the time that he fell ill was doing much the same thing his ancestors had done for generations before him, and who may not even have been aware that there was a war on. The same is true if it started on a farm in Kansas. Only if the origin was French can the pandemic truly be described as an outcome of the conflict, because in that case it was brewed in a camp where men were brought together (with some women) for the express purpose of killing other men. There is one final possibility: that none of the three theories is correct, and the real origin of the pandemic has yet to be proposed.

12

Counting the dead

How many had died? People wanted to know from the moment it was over, not only to gauge the pandemic's impact on humanity, and to set the historical record straight, but also to extract lessons from it for the future. They had an idea of the scale of the previous flu pandemic, the Russian flu of the 1890s. It had killed around a million people. If the Spanish flu were in that ballpark, then perhaps a flu pandemic was simply something that happened periodically, and one had to learn how to manage it. If it were much larger, however, the conclusion would have to be different: something about that particular flu, or about the state of the world in 1918, or both, had created a deadly anomaly.

In the 1920s, an American bacteriologist named Edwin Jordan estimated that 21.6 million people had died from the Spanish flu. Right from the beginning, therefore, it was clear that it was in a league of its own. This was higher than the death toll of the First World War, and *twenty times* higher than the death toll of the Russian flu. We now know that Jordan's figure was an underestimate, but it was one that stuck for close to seventy years, meaning that for a long time after the event, the human species had only a tiny inkling of its loss.

Jordan can be forgiven. Epidemiology was young in 1920. Diagnostic criteria for influenza and pneumonia were vague, and many countries didn't count deaths in peacetime, let alone in the midst of a boundary-shifting, chaos-generating war. Where data were available, he could calculate excess mortality rates – a measure of the number of people who died over and above what might have been expected in a 'normal' or non-pandemic year – but these hid a multitude of diagnostic sins. There was no such thing as a 'laboratory confirmed death' from flu in 1918, because nobody knew that flu was caused by a virus. What's more, flu pandemics don't really start or stop. They invade the seasonal flu cycle, grotesquely distorting its morbidity (sickness) and mortality (death) curves, then recede until those curves reveal themselves again. Even now that the tools exist to differentiate seasonal and pandemic strains, defining a pandemic's limits is an essentially arbitrary task.

In 1991, two American epidemiologists, David Patterson and Gerald Pyle, raised Jordan's bid to 30 million – hinting at a bigger disaster, though still not one on the scale of the Second World War, which eliminated roughly twice as many souls. They incorporated new data that had come to light since Jordan's day, but they only counted the death toll from the second, autumn wave. There were some areas of the world for which they had no better data than Jordan. They echoed his estimate of 450,000 dead for Russia, for example, along with his caveat that it was no more than 'a shot in the dark'. 'Little is known about the toll in China,' they wrote, 'but with some 400–475 million inhabitants the loss of life could have been enormous.'[1] Russia and China were big, populous countries. Errors in the calculation of their death tolls would have a serious impact on any global tally, so it is worth examining Patterson and Pyle's estimates for them in a little more detail.

The estimate of 450,000 deaths corresponds to roughly 0.2 per cent of the Russian population at that time. If that were correct, then Russia suffered the lowest flu-related mortality in Europe, which seems counter-intuitive in a country in the grip of a civil war, where the infrastructure of daily life had broken down. The Odessan case suggests, indeed, that it was not correct, and that the real number may have been higher. We know that Odessans were often infected with more than one disease at a time, and that the chances of misdiagnosis were high. Tizengausen, the pathologist at the Old City Hospital, found lung haemorrhage, a telltale sign of Spanish flu, in a larger number of corpses than had been diagnosed with it while alive. Tizengausen had a second job at the city morgue, and there he found the same signs in cases that had been wrongly diagnosed as cholera or, more vaguely, 'plague'. He also discovered that some of those who had been correctly diagnosed with Spanish flu had been infected simultaneously with typhoid, dysentery, TB and other serious diseases.

Vyacheslav Stefansky, Yakov Bardakh's former student who also worked at the Old City Hospital, noted that around 8 per cent of flu patients who were admitted to his hospital went on to die of the disease, and another doctor recorded a similar proportion at the Jewish Hospital. This compares to a global case fatality rate of 2.5 per cent.[2] In the 1950s, a team of Russian epidemiologists led by V. M. Zhdanov estimated that 70,000 Odessans were sick with the Spanish flu in October 1918.[3] If they were right, and if the case fatality rates calculated by Stefansky and his colleague at the Jewish Hospital were right, then around 6,000 Odessans died of *ispanka* in that month. That equates to 1.2 per cent of the population, or six times Patterson's and Pyle's estimated mortality rate for the country and the autumn wave as a whole.

Zhdanov felt that Odessa had suffered worse than any other major Russian city, so if Russia had been composed exclusively of cities, then we might have to revise that figure downwards. But Russia wasn't composed exclusively of cities, of course. Urban folk were, in fact, very much in the minority, accounting for somewhere between 10 and 20 per cent of the population. And if the flu was bad in Odessa, it probably wasn't any better in the surrounding countryside, where it was not uncommon for tens of thousands of people to depend on a single doctor with no drugs at his disposal. As we've seen, drugs didn't work. But doctors themselves – and more importantly, nurses – *could* make a difference, and they were signally lacking. When the International Committee of the Red Cross sent French officer Ernest Léderrey to inspect the sanitary situation in Ukraine in 1919, he reported that some villages had lost 10 to 15 per cent of their inhabitants to typhus and Spanish flu the previous year, and dysentery had added to their woes (doctors noticed that Spanish flu often finished off the starving). With the onset of winter, what was left of the *zemstvos* – pre-revolutionary provincial councils – had tried to help by setting up temporary hospitals. 'But what are fifty or sixty beds, when every house has at least one invalid who should be isolated,' wrote Léderrey, 'A drop in the ocean!'[4] If we apply the 1.2 per cent mortality rate to the country as a whole, 2.7 million Russians died of Spanish flu.

China remains a conundrum, mainly because it is impossible to define the Chinese epidemic. In a country where a year did not pass without an epidemic, it is possible that Spanish flu came sandwiched between two visitations of pneumonic plague, that all three waves of respiratory disease – those of December 1917, October 1918 and December 1918 – were caused by the influenza virus, or that another, as yet unidentified microbe was responsible for one or more of them.

America and Britain, wealthy countries, lost approximately 0.5 per cent of their populations to the Spanish flu. Extrapolating from poorer countries, but assuming that China suffered less badly than India (where the rate was ten times that in America), Patterson and Pyle came up with a range of between 4 million and 9.5 million deaths in China. But they had no Chinese data to work from, because there was no centralised collection of health data in China during the warlord period, and the missionaries who rode to the rescue of the ailing did not gather statistics systematically. The only parts of the country where some health statistics were gathered as a matter of routine were those that were under foreign control, and in 1998, a Japanese scholar, Wataru Iijima, made use of these to come up with a new estimate. Basing his calculations on foreign-controlled Hong Kong and southern Manchuria, and with many caveats, he estimated that only a million Chinese people died.[5]

Iijima's estimate is problematic, however. One of the assumptions he made was that the flu arrived at the ports, and that poor communications prevented it from penetrating the interior. Yet Taiyuan, the 'capital' of Shansi and very much inside that interior, was already connected to Peking by rail in 1918, and anecdotal evidence suggests that the epidemic was anything but mild in Shansi. In 1919, a man who had first-hand experience of fatal epidemics, Percy Watson, described the outbreak there – by which he meant the illness that raged over three weeks in October 1918 – as 'one of the most fatal epidemics reported in the medical literature this past year'.[6] On 2 November 1918, describing the same outbreak, the *North China Herald* mentioned thousands of dead in Taigu, a town in Shansi. And contemporary reports kept by the Chinese Post Office referred to many victims in two neighbouring provinces, Hopei to the east and the confusingly named Shensi (Shaanxi) to the west. In Hopei, the flu was reported to have killed more postal workers than had a visitation of pneumonic plague early in 1918. It seems at least possible, therefore, that flu was widespread in China in 1918 and 1919, that it followed a similar pattern as elsewhere in the world – of mild spring wave, severe autumn wave and possible recrudescence in early 1919 – and that in parts of the country, at least, the death toll was very high indeed. In the case of China, Patterson and Pyle may have been closer to the mark.

In 1998, on the eightieth anniversary of the pandemic, Australian historian and geographer Niall Johnson and German flu historian Jürgen Müller revised the global death toll upwards again. Their justification was that the earlier estimates represented tips of a largely unreported iceberg, that the under-reporting affected rural populations and ethnic minorities disproportionately, and that there were indications that some of those populations – partly for reasons of historical isolation – had suffered very heavy losses. By then, the death toll in India alone had been estimated to be as high as 18 million – three times what Indians believed it to have been in 1919 – making Jordan's 21.6 million seem 'ludicrously low' by comparison. Johnson and Müller came up with a figure of 50 million, of which Asia accounted for 30 million. But, they stressed, 'even this vast figure may be substantially lower than the real toll, perhaps as much as 100 per cent understated'.[7]

An understatement of 100 per cent means that the number of dead could have been as high as 100 million – a number so big and so round that it seems to glide past any notion of human suffering without even snagging on it. It's not possible to imagine the misery contained within that train of zeroes. All we can do is compare it to other trains of zeroes – notably, the death tolls of the First and Second World Wars – and by reducing the problem to one of maths, conclude that it might have been the greatest demographic disaster of the twentieth century, possibly of any century.

In the annals of flu pandemics, the Spanish flu was therefore unique. Most scientists now agree that the event that triggered it – the spillover of the pandemic strain from birds to humans – would have happened whether or not the world had been at war, but that the war contributed to its exceptional virulence, while at the same time helping to spread the virus around the world. It would be hard to think of a more effective dissemination mechanism than the demobilisation of large numbers of troops in the thick of the autumn wave, who then travelled to the four corners of the globe where they were greeted by ecstatic homecoming parties. What the Spanish flu taught us, in essence, is that another flu pandemic is inevitable, but whether it kills 10 million or 100 million will be determined by the world into which it emerges.

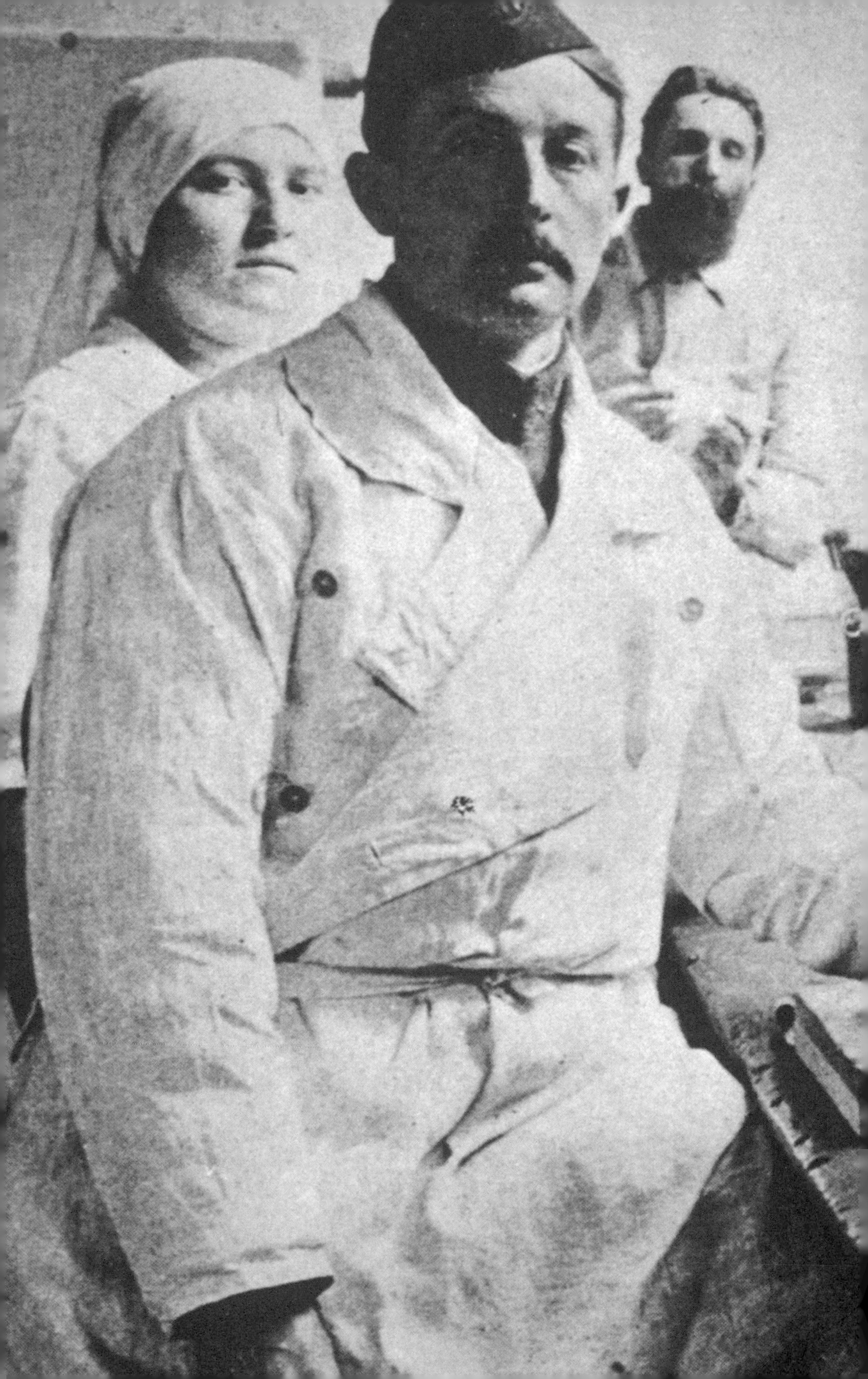

PART SIX: Science Redeemed

René Dujarric de la Rivière in an army laboratory, Calais, 1915

13

Aenigmoplasma influenzae

In the dog days of August 1914, an ageing Ilya Mechnikov – Russian exile, Nobel laureate, 'lieutenant' of Louis Pasteur and mentor of Yakov Bardakh, Wu Lien-teh and others – battled his way across a Paris in the grip of mobilisation to reach the Pasteur Institute, one of the world's leading centres for the study of infectious diseases and the production of vaccines. When he arrived, he found it under military command. Most of the younger scientists had left for active service and all of the experimental animals had been killed. The man who had renounced God at the age of eight, who believed fervently that the progress of civilisation depended on the advancement of science, surveyed his deserted empire and quaked.

In his novel *Journey to the End of the Night*, Louis-Ferdinand Céline immortalised Mechnikov as Serge Parapine, an eccentric and demented genius who 'always had enough hair on his cheeks to make him look like an escaped convict', and who raged and muttered through the smelly corridors of the renowned Parisian institute where he worked. The institute's other inhabitants were 'grey-haired, umbrella-carrying schoolboys, stupefied by the pedantic routine and intensely revolting experiments, riveted by starvation wages for their whole adult lives to these little microbe kitchens, there to spend interminable days warming up mixtures of vegetable scrapings, asphyxiated guinea pigs, and other nondescript garbage'. But as Mechnikov intuited on that summer's day, the era that Céline described so scathingly – his era, in which important battles had nevertheless been won against crowd diseases, and faith in science rode high – was about to end.

First, though, there was a war to be fought – and diseases to be kept at bay. One of the young scientists who had left the Pasteur Institute at the outbreak of war was René Dujarric de la Rivière, a twenty-nine-year-old aristocrat from the Périgord who, like others of his contemporaries, had been swallowed up by the army's network of laboratories. Four years later, when the second wave of the Spanish flu broke out, he was working in the central army laboratory in the city of Troyes. 'I was there in the Champagne region when an artillery troop came through on its way to the front. They never left. All of them, men and officers alike, were suddenly struck down and had to be hospitalised urgently.'[1] The army launched a vaccination campaign, using a vaccine against pneumonia-causing bacteria that had been developed at the Pasteur Institute before the pandemic. Dujarric had spent time in Richard Pfeiffer's lab in Breslau – where Pfeiffer, known to his colleagues as the 'Geheimrat' or privy counsellor, was treated with profound respect – but he had begun to doubt that Pfeiffer's bacillus was really the cause of flu.

He wasn't alone. Pfeiffer's bacillus – *Haemophilus influenzae*, to give it its scientific name – is a real bacterium that lodges in the nose and throat and causes infections, some of them severe, but while it had been found in many of the flu cases analysed, it hadn't been found in them all. In New York, bacteriologists William Park and Anna Williams of the city health department had collected lung tissue from dozens of flu victims post-mortem, then grown the bacteria colonising it on agar gel in order to identify the species present. Even when Pfeiffer's bacillus was among them, they found, it seemed to exist in different strains. That was odd: in a pandemic, you'd expect to find the same strain consistently. And it certainly wasn't the only bacterium in the mix: streptococci, staphylococci and pneumococci were there too, in legions, and they could also cause respiratory disease. Alexander Fleming, a captain in the British Army's medical corps at the time, had confirmed Park's and Williams' results using tissue from, among other places, Étaples. Some had gone a step further. As early as 1916, Milton

Rosenau, a doctor in Boston, had voiced his suspicion that the causative agent of flu was a virus – an organism small enough to pass through the pores of the porcelain Chamberland filters that were routinely used to trap bacteria out of liquid at that time, and hence commonly referred to as a 'filterable virus'.

Dujarric probably knew of Fleming's work, possibly even of Park's and Williams', and of Rosenau's suspicions. In 1915, before moving to Troyes, he had run the army laboratory for the northern region, in Calais, and while there had crossed paths with Sir Almroth Wright, the British inventor of the typhoid vaccine. Wright had requisitioned the casino in nearby Boulogne for a laboratory – beds replaced gaming tables, chandeliers were swathed in linen sheets – and put his junior colleague Fleming and others to work in it. They shared the space with an American hospital set up by Harvard University. Wright was well known by then, and the casino received a constant stream of visitors. He got on well with the French, according to Fleming's French biographer, André Maurois (who acted as an interpreter and liaison officer with the British Army), despite differences in the British and French attitudes to the war. For the French it was a quasi-religious ceremony to be treated with great solemnity, while the British did their duty and took what opportunities they could for relaxation. Maurois recounts how Fleming and another man, probably Wright, were enjoying a wrestling match one day, when a door opened and in came a delegation of senior French Army doctors. The wrestlers leapt to their feet and immediately engaged the visitors in a scientific discussion, but, recalled a witness, 'I will never forget the expression on the French doctors' faces on discovering that scene.'

They may not have agreed about the place of contact sports in the theatre of war, but they were converging on the notion that the cause of flu might not be Pfeiffer's bacillus. That idea was therefore in the air when, walking in the streets of Troyes one day, early in October 1918, Dujarric bumped into his old friend and fellow Pasteurian, Antoine Lacassagne. The two had not seen each other since before the war, but Lacassagne had been sent to Troyes to help vaccinate the troops. 'After chatting a moment, he made me a curious proposition,' Lacassagne recalled years later. 'Dujarric asked me to do him the favour of injecting him with the filtered [blood] of a flu patient, the experiment that he felt would confirm his hypothesis. I pointed out the moral dilemma he was placing me in, but he finally convinced me that it was better that I do it, in the best conditions, than that he inject himself – something he was otherwise determined to do. I administered the injection on the morning of Tuesday 8 October, in his army laboratory.'[2]

Lacassagne had to leave for Paris the next day, and he didn't discover the outcome of the experiment until months later. For two days, Dujarric remained well, then he noticed the first symptoms. He managed to describe the course of the disease: 'Third and fourth day, after an abrupt onset, intense and persistent frontal headache, pain all over . . . temperature between 37.8° and 38.2° . . . Fourth night agitated, nightmares, sweats. On the fifth day the pain disappeared; very pleasant euphoria after the indefinable sense of malaise that had marked the previous two days . . . In the following days everything returned to normal, except for a lingering fatigue, then on the seventh day cardiac symptoms emerged, and these persist: intermittent but very disagreeable chest pains, irregular pulse, breathlessness at the slightest effort.'

In a second experiment performed a few days later, he painted his own throat with a filtered emulsion of flu patients' sputum and waited, but experiencing no further symptoms, concluded that the first experiment had immunised him against the second. Miraculously, given his own state of health and the chaos around him, he managed to write up his findings and transmit them to the Pasteur Institute's director, Émile Roux, within a matter of days. It was only a preliminary study, he admitted in the report that Roux presented on his behalf on 21 October, to the French Academy of Sciences, but the key point was that the blood with which he had been injected had been filtered, hence free of bacteria. It raised the possibility that the flu was caused by a virus.[3]

What did Dujarric mean by a virus? He probably wasn't quite sure himself. All he could really say was that it was something smaller than a bacterium, that was capable of transmitting disease. He probably would have hesitated before describing it as a living organism, however (and indeed, the debate over whether a virus is dead or alive continues today: can an organism be described as alive if it is incapable of reproducing on its own?), and he may have at least allowed the possibility that what he had infected himself with was something more like a venom.

Coincidentally, in the same proceedings of the academy, two other Pasteurians, Charles Nicolle and Charles Lebailly, reported the same conclusion. They were working in the Pasteur Institute's outstation in Tunis, and in the first days of September they had inoculated a monkey and two human volunteers with the sputum of a Spanish-flu patient – unfiltered in the case of the monkey, filtered in the case of the humans. The monkey, which had received the inoculum via the inner lining of its eyelids and nostrils (considered a part of the airborne route), showed signs of a flu-like disease a few days later – high temperature, loss of appetite, lassitude. The human who had received the filtrate under his skin fell sick on the same day, but the one who received it into his blood remained well. Nicolle and Lebailly concluded that the cause of the disease was a filterable virus that could not be transmitted by the blood.

Dujarric de la Rivière and Nicolle and Lebailly were the first to publish, independently but simultaneously, the finding that the flu was probably caused by a virus. Before 1918 was out, German, Japanese and British scientists had performed similar experiments and arrived at similar conclusions. Like Dujarric, the German, Hugo Selter of the University of Königsberg, had experimented on himself. The first half of the twentieth century was an era of self-experimentation (Mechnikov had deliberately given himself cholera, among other potentially lethal diseases) but perhaps it was easier to risk one's life when all around you were risking theirs – that is, in time of war. Members of the British team, who published an initial account of their findings in December 1918, did not experiment on themselves. But one of them, Graeme Gibson, was preparing a follow-up report when, worn down by long hours in the army lab at Abbeville, near Étaples, he caught the flu. He died before it was published the following March.

For all their bravery, the credibility of these scientists' findings is tainted. The experiments were conducted during the pandemic, at a time when it would have been impossible for them to ensure that their laboratories were free of contamination by the ubiquitous flu virus, so it is hard to know by which route their experimental subjects received the infection. Anyone paying attention will have noticed that Dujarric's and Nicolle's and Lebailly's results contradict each other: Dujarric thought that he had given himself flu via an injection of filtrate into his blood, while the pair in Tunis ruled out the blood as a transmission route. Nicolle and Lebailly were right, in fact: influenza is not transmissible by the blood, so Dujarric cannot have caught it from the injection that Lacassagne gave him. He probably caught it via the usual route – the air – while bending over the four gravely ill soldiers whose blood he took in preparation for the experiment, developing symptoms after the usual incubation period of two or three days. As happens so often in science, in other words, Dujarric was right for the wrong reasons.

Rosenau and his colleague John Keegan in Boston also tried, in the thick of the autumn wave, to demonstrate that the causative agent of the flu was filterable, but they were unable to transmit the disease. Others failed too, but their results are as unreliable as those of their French counterparts. One reason their human volunteers may have failed to get sick, for example, was that they had been exposed to the virus during the spring wave, and acquired some immunity. Within the scientific community at the time, however, people interpreted the results according to their preferred theory. The Geheimrat himself, Richard Pfeiffer, remained convinced that 'his' bacillus was the most likely candidate. His supporters felt that if Rosenau had found no virus, it was because there was no virus to be found (putting his trust in the data, Rosenau agreed with them – a case of being wrong for the right reasons). When it came to explaining the troubling finding that Pfeiffer's bacillus was absent from the lungs of some flu victims, on the other hand, the Pfeiffer camp blamed bad tools and methods. It didn't help, in terms of penetrating the shadows around the disease, that antibacterial vaccines had shown some efficacy against it – because they had worked against those lethal secondary infections.

Only in the 1930s did the shadows begin to lift. One of the peculiar aspects of the 1918 pandemic was that it coincided with an epidemic of a very similar disease in pigs – so similar, in fact, that the pig disease was dubbed 'swine flu'. At the time, veterinarians regarded it as a new disease in swine, but from then on it erupted periodically in herds. In 1931, following one such outbreak, an American virologist, Richard Shope, confirmed what Dujarric, Selter and the others had tried to demonstrate earlier, in far more difficult circumstances: that flu was caused by a filterable virus. Two years later, a team of British scientists working at the National Institute for Medical Research in London, did the same in humans. After a ferret sneezed in the face of one of them, Wilson Smith, he came down with flu. They went on to show that a filterable agent could transmit flu from a ferret to a human and back again (whether that agent was an organism or a toxin was still an open question, though by 1950 the London team had come to believe, correctly, that they were dealing with an organism).

From the humble beginnings of a ferret's sneeze, the vast and complex biology of influenza began to unfold. When a virus infects a person, his or her immune cells secrete tiny morsels of protein called antibodies that attach themselves to the virus, disabling it. Antibodies can linger in the blood for years after the infection has passed, providing a record of past infections, and by the 1930s, scientists already had tests for detecting them in serum (the clear liquid in which all the other components of blood float).

When they saw that antibodies produced during one flu outbreak did not necessarily protect people against another, they realised that flu came in different varieties. Three types of flu were eventually identified (a fourth has been added very recently): A, B and C. A and B cause epidemics, but only A causes pandemics. C is altogether milder and less contagious than the other two. The virus that caused the Spanish flu was, needless to say, an A.

One of the reasons that scientists had such difficulty believing that a virus had caused the pandemic was that, unlike many of the opportunistic bacteria that invaded the lungs of patients already infected by flu, and no matter what nourishing gel they offered it, they weren't able to grow it in a dish. 'Grow', in this sense, means persuade it to produce more copies of itself. As we know, however, a virus can't reproduce outside a host cell. For a virus to enter a host cell, protein structures on its surface, called antigens, must first bind to receptors on the cell's surface. The two have to fit pretty well, like a lock and key, but when they do, a cascade of molecular events is unleashed that allows the virus to pass inside the cell (an antibody works by attaching itself to one of those same antigens, so preventing it from attaching itself to the host receptor). Once inside the cell, the virus requisitions the cell's reproductive machinery to manufacture new copies of its components. These are then assembled into new viruses that break out of the cell, killing it in the process, and proceed to infect new cells. In humans, the flu virus invades cells lining the respiratory tract, damaging that lining as the cells die. The result is the symptoms of flu.

In 1931, the same year that Shope identified a virus as the cause of swine flu, American pathologists Alice Woodruff and Ernest Goodpasture succeeded in growing a virus in a fertilised chicken's egg. This was the result of their observation that chicken eggs could become infected by a disease of poultry called fowlpox, which is caused by a virus. Their achievement meant that viruses could now be grown in large quantities in the laboratory, free of contamination by bacteria. That in turn meant that scientists could study them tranquilly, outside epidemics, and begin developing vaccines against them. The first flu vaccine was produced by a Russian, A. A. Smorodintseff, in 1936. He took a flu virus and grew it in an egg, then extracted the offspring of the virus that replicated least well and grew them in another egg. He repeated this process thirty times, until he had a virus that didn't replicate very well at all – another way of saying that it was mild – and this he injected into people. The first human guinea pigs experienced a barely perceptible fever but were protected against reinfection with flu.

Smorodintseff's vaccine was given to Russian factory workers with the goal of reducing absenteeism due to respiratory disease. The same kind of vaccine was used for the next fifty years in the Soviet Union, and more than a billion Russians received it. But it only protected against influenza A, and it had other limitations too – not least that the virus could continue to reproduce in the recipient and potentially recover its virulence. Later, scientists found that they could stop it replicating by treating it with the chemical formaldehyde. Though much larger quantities of it were required, the 'inactivated' virus still provided protection against reinfection.

Vaccines were developed that protected against more than one flu type – so-called polyvalent vaccines – and by 1944 the American troops arriving in Europe to fight the Second World War had received the first mass flu vaccine containing inactivated viruses of more than one type. One of those who had worked on it was Jonas Salk, the man who would become famous as the inventor of the polio vaccine (and whose name, in the 1950s, would be better known to Americans than that of their president). His fascination with viruses began in the early twentieth century, when the world's virologists – some of them in his native New York – were trying to solve the mystery of the Spanish flu.

By the 1940s, therefore, scientists had classified flu, they had introduced it to all manner of unsuspecting animals, they had even – in a tribute to human ingenuity – developed vaccines against it. But even after all doubts had been silenced as to the existence of the flu virus, it remained a mythical beast – something like a leprechaun, or the Higgs boson before it was outed in 2012 – because nobody had ever seen it. It belonged to that category of creature that, in a premonitory article written in 1903, Émile Roux had labelled *êtres de raison*, or theoretical beings: organisms whose existence can be deduced from their effects, though they have never been detected directly.[4]

The problem was that, even with the help of an optical microscope, there was a limit to the tininess of objects that could be seen. Essentially, it was impossible to see anything smaller than the wavelength of visible light. Red blood cells were visible, as were some of the bacteria that infected them, but not a virus, which is smaller. Two Germans, Max Knoll and Ernst Ruska, broke through that barrier in the early 1930s, when they invented the electron microscope. An electron, like a photon of light, behaves as both wave and particle, but its wavelength is hundreds of times shorter than that of a photon. The flu virus was visualised for the first time in 1943, twenty-five years after Dujarric de la Rivière risked his life to prove it existed.

It is of medium size, as viruses go, and close to spherical (though sometimes it can be shaped like a rod): a tiny bead of protein encircling an even tinier kernel of genetic information. The whole is enclosed by a membrane, on top of which sits that all-important antigen, called haemagglutinin, or H for short. H looks like a lollypop. Its stalk projects down into the membrane while its round if convoluted head is presented to the outside. In fact, some flu viruses – including the influenza A viruses that cause pandemics – carry not one but two major antigens on their surface. H is the metaphorical crowbar that allows the virus to break into a cell, while neuraminidase (N), the second major antigen, is the glass cutter that allows it to exit again.

Flu's genetic material consists of single-stranded RNA, as opposed to double-stranded DNA in humans, and this RNA is packaged into eight segments (for ease, we'll call them genes). Two of these genes are translated into the surface proteins H and

N, while the other six – the so-called internal genes – encode proteins that modulate functions such as the virus's ability to replicate or to fend off the host's immune response. When the flu virus reproduces itself, these genes have to be copied, but because RNA is less chemically stable than DNA, the copying mechanism is sloppy, and errors creep in. This sloppiness is the key to flu's notorious lability – that capacity it has to generate endless new variations on itself – because errors at the genetic level translate into structural changes in the proteins they encode, and even tiny ones can have a big effect. Every year, for example, about 2 per cent of the units – called amino acids – that make up flu's surface proteins are replaced. That's enough to alter the shape of the H antigen such that an antibody that once bound to it can no longer do so very well. The virus 'escapes' the host's immunity, partially, and causes a new, seasonal outbreak. It is the reason why flu vaccines have to be updated each year.

That slow accumulation of errors is known as drift, but flu can also reinvent itself in a more radical way. This happens when two different flu viruses meet in a single host, swap genes and produce a new one – a virus with a novel H-N combination, for example. This kind of change, called shift – or more memorably, 'viral sex' – tends to trigger a pandemic, because a radically different virus demands a radically different immune response, and that takes time to mobilise. If the two 'parent' viruses come from two different hosts – a human and a bird, say – their encounter may result in an antigen that is novel to humans being introduced into an otherwise human-adapted virus. Every flu pandemic of the twentieth century was triggered by the emergence of a new H in influenza A: H1 in 1918, H2 in 1957 and H3 in 1968.

Once the human immune system has been mobilised against the new virus, it enters a more stable equilibrium with its host. The pandemic passes, but the virus continues to circulate in a benign, seasonal form, provoking occasional outbreaks as it evolves through drift. That equilibrium is maintained until another novel virus emerges. But an old H can also cause a new pandemic, if it emerges in a population that has become immunologically naive again – that is, in a generation that has never been exposed to it. In other words, it can be recycled over roughly the human lifespan. There is some evidence that H3, which caused the 'Hong Kong' flu pandemic of 1968, also caused the Russian flu of the 1890s, while H1 caused the Spanish flu of 1918 and the so-called 'swine flu' (actually a human flu) of 2009. A novel N may also be capable of triggering a pandemic (this is currently a subject of debate), and there are, to date, eighteen known varieties of H and eleven known varieties of N. Nowadays, therefore, influenza A viruses are classified by subtype according to which versions of these two antigens they carry. A given subtype can be further divided into strains, depending on the make-up of its internal genes. The subtype that caused the Spanish flu was H1N1 – all the ones, a ghostly echo of 'disease eleven', as French Army doctors dubbed it, on the far side of a gulf of knowledge.

14

Beware the barnyard

Extinct until 2005, the H1N1 strain that caused the Spanish flu is, today, alive and well (if we can call a virus alive) and imprisoned in a high-security containment facility in Atlanta, Georgia. It was brought back to life for the purposes of scientific study, though not everybody was persuaded of the wisdom of that move. Fellow scientists accused those responsible of having revived 'perhaps the most effective bioweapons agent now known'. Since the method for its reconstruction was available on the Internet, they argued, 'its production by rogue scientists is now a real possibility'.[1]

The researchers who reanimated the virus (two groups, to date) countered that doing so would help them answer critical questions about what happened in 1918, and so prevent a similar disaster from happening again. The virus remains safely tucked away in its level-four biohazard laboratory, nobody has unleashed it on the world, and it has indeed shed light on the 1918 pandemic – so that, for now, the cost-benefit analysis seems to be in favour of those who revived it.[2]

By the 1990s, there were still many outstanding questions about the Spanish flu. Of all the flu pandemics in living memory, and even some that we know about only through historical texts, it was the odd one out. It was the most deadly. Although the vast majority of its victims experienced something not much worse than seasonal flu, it killed a much higher proportion of them – at least 2.5 per cent, compared to less than 0.1 per cent for other flu pandemics (making it at least twenty-five times as lethal). It was vicious in its own right, and it was also more likely to be complicated by pneumonia, which was usually the ultimate cause of death. Its mortality 'curve' was W-shaped, not U-shaped as is typical of flu, with adults aged between twenty and forty being particularly vulnerable, as well as the very young and the very old. It seemed to strike in three waves, but the first two waves presented so differently – the first being confused with seasonal flu, the second with pneumonic plague – that many people doubted they were caused by the same organism (the third wave, which was intermediate in virulence between the other two, aroused less curiosity). Whereas previous flu pandemics had tended to take three years to circle the globe, this one raced around it in two at the outside. And finally, it wasn't clear where it had come from. Origins in France, China and the US would all be proposed.

The only thing that most people agreed upon was that it probably had its origins in birds. Wild waterbirds had been considered the natural reservoir of influenza A since the 1970s, when an American veterinarian named Richard Slemons isolated the virus from a wild duck.[3] His discovery motivated others to conduct surveys of wild bird populations, and thanks to their efforts we know that waterbirds harbour a vast diversity of flu – not in their lungs as in humans, but in their digestive tracts, and generally without suffering any ill effects. They shed the virus into the water in their droppings, from where other birds pick it up, and different strains meeting in the same bird may swap genes to produce a novel one. Ducks make particularly good flu incubators. When, soon after Slemons' discovery, French virologist Claude Hannoun surveyed five species of migratory duck in the Somme estuary, he found that they harboured around a hundred different flu strains between them. Often, an individual bird harboured more than one strain, and some were hybrids that didn't match any known subtypes. Hannoun had caught flu, in other words, in flagrante delicto – in the act of evolving.[4]

In the 1990s, however, nobody suspected that a bird flu virus could infect a human or cause a pandemic. The receptors on a cell lining a human lung are shaped differently from those on a cell lining a duck's intestine, and the prevailing idea was that, for the virus to jump to humans, an intermediate host was required, in which it could adapt from one receptor type to another. That intermediate host was thought to be pigs. The cells lining a pig's respiratory tract carry receptors to which both bird and human flu viruses can bind, meaning that pigs provide an ideal crucible for the mixing of a novel strain that infects people.

Following this line of thought, John Oxford, the man who proposed a French origin for the Spanish flu, pointed out that Étaples was only fifty kilometres from the Somme estuary – a major stopping point on the route taken by waterbirds migrating from the Arctic to Africa – and that the camp had its own piggery. Camp caterers brought in live poultry that they had purchased in surrounding villages, and some of these domesticated birds may have been infected, having mingled with wild birds passing through the bay. (By way of comparison, Haskell County, the putative Kansas origin of the pandemic, is 200 kilometres from the nearest significant wetlands, Cheyenne Bottoms in Barton County, while the closest wetlands to Taiyuan, the capital of Shansi, are 500 kilometres away and beyond the provincial borders.) It was only the death in 1997 of a little boy in Hong Kong, from a flu subtype that was known in birds but had not previously been detected in humans – H5N1 – that raised the frightening possibility that a flu virus could be transmitted directly from birds to humans. At that point the question had to be asked: could this also have happened in 1918?

By the 1990s, gene sequencing had become a powerful tool, and scientists had begun to hope that it might help them solve the puzzle of the Spanish flu. A gene consists of thousands of units called bases. If they could determine the sequence of those units across all eight genes of the Spanish flu virus, and compare it to those of other flu viruses, perhaps they would discover why that pandemic was so unusual. Unfortunately, by the 1990s, the Spanish flu was a distant memory, so the first challenge was to obtain a sample of the virus. That meant finding infected lung tissue that had been preserved for nearly eighty years, and it wasn't just the tissue that had to have survived, but the records that went with it. The race was on: pathologists began scouring the planet for the elusive microbe.

The first glimmer of success came in 1996, when biologist Ann Reid and pathologist Jeffery Taubenberger discovered it hiding in almost plain sight, at the US Armed Forces Institute of Pathology (AFIP) in Washington DC where they worked. It was in a scrap of lung that had been stored there ever since an army pathologist had removed it from Roscoe Vaughan, a twenty-one-year-old private who had died at a military camp in South Carolina, in September 1918. The tissue had been treated with formaldehyde to preserve it, and embedded in paraffin wax. The formaldehyde had damaged the virus's RNA, so the scientists were only able to sequence fragments of it (they later obtained a second flu-containing sample from AFIP), but they published these first partial sequences in 1997, and a doctor in San Francisco named Johan Hultin happened to read their paper.

Hultin, who was in his seventies by then, had a long-standing interest in the Spanish flu. In 1951, as an eager young medical student, he had set out to find the virus himself. He knew that there were places in Alaska where people had died in large numbers, and been buried in mass graves, and he thought that if the permafrost had preserved them, he might be able to extract the virus from their remains. He organised an expedition to the village of Brevig Mission on the Seward Peninsula (about 800 kilometres north of Dillingham), which lost 85 per cent of its population in five days in 1918, and having obtained permission from the village council, excavated the grave where the victims had been buried. He found lung tissue and brought it back with him, intending to analyse it in the lab. But it was 1951. Though scientists knew viruses existed, though they had seen them under electron microscopes and grown them in eggs, they couldn't extract the fragile organisms from decades-old tissue that had – despite the misleading term 'permafrost' – been through cycles of potentially damaging freezing and melting. Hultin shelved the project and moved on to other things.

Nearly five decades later, he barely skipped a beat. Back he went, alone, to the same mass grave. This time, he discovered the remains of a woman who had been overweight in life, so that the fat around her torso had protected her lungs from the worst ravages of decomposition. He packaged up her lung tissue, posted it to Taubenberger, replaced two large crosses that had marked the grave in 1951 and since rotted away, and got back on the plane to San Francisco. Taubenberger succeeded in extracting viral RNA from the tissue – though this too had been damaged, in this case by the freezing-melting cycle – and to sequence further fragments. In 2005, after nine years of painstakingly 'stitching' the partial sequences together, he and Reid published the first complete sequence of the Spanish flu virus (Taubenberger's group has since repeated that feat in a couple of weeks, using a new, high-powered sequencing technique). Further partial sequences were obtained from samples stored in hospital archives in London.

The first thing Reid and Taubenberger noticed about the sequence was how very like known sequences of bird flu it was. The virus had kept much of its bird flu-like structure, which might explain why it was so virulent: it was a very alien invader that took the human immune system by storm in 1918, yet one that could still recognise – that is, bind to – human cells. It was, in other words, a formidable vehicle of disease. Reconstructing it was a natural next step, though they deliberated long and hard about it. With virologist Terrence Tumpey and others at the Centers for Disease Control and Prevention (CDC) in Atlanta, they 'fed' the viral sequence to human kidney cells growing in a dish, forcing them to manufacture the virus just as a virus forces a host cell to do in the normal process of infection. Then they infected mice with it, and saw just how formidable it really was.

The main sign of flu infection in mice is loss of appetite and weight. Two days after Tumpey's team infected the mice with the revived virus, they had lost 13 per cent of their body weight.

Four days after infection, they had nearly 40,000 times as many viral particles in their lungs as mice infected with a seasonal strain. And six days after infection, they were all dead, while the control mice were still standing. Mice aren't humans, nevertheless the contrast was dramatic.

When a virus invades the human body, the body's immune system is spurred into action. Within minutes, immune cells start secreting a substance called interferon that blocks the synthesis of new protein, so arresting the production of new viruses. But after millennia of co-evolution with humans, flu has evolved its own means of blocking interferon. It does so by concealing the evidence that it has hijacked the cell's reproductive machinery, so that interferon can't shut it down. Taubenberger's team found that the 1918 virus was exceptionally good at this, giving it a head start when it came to replication.

Interferon is the body's first line of defence, a generalised rapid response to invasion that is deployed while the immune system musters a rebuff that is more tailored to the invader in question. If interferon works, the invasion is halted and the individual barely feels unwell. If it fails, it means the virus has been able to replicate, and the body's second line of defence is mobilised. Antibodies and immune cells converge on the site of infection. The immune cells release chemicals called cytokines that, among other things, increase blood flow to affected tissues so that more immune cells can reach them. They also kill other host cells, if necessary, to stop the infection spreading. The result is redness, heat, swelling and pain – collectively known as inflammation.

Inflammation on a massive scale is what the world's pathologists saw in 1918 – those red, engorged lungs that were hard to the touch and seeped a watery, bloody fluid. Rereading their reports, immunologists from the 1940s on thought that those pathologists had witnessed the effects of a 'cytokine storm', an overzealous, second-line immune response that ultimately caused more damage than the virus it was intended to destroy. This is what Taubenberger and his colleagues saw in animals infected with the resurrected virus too. Whereas a benign, seasonal virus produced a transient cytokine response and localised, superficial damage to the lung, the 1918 variety produced a strong, prolonged cytokine response and damage that was severe and deep. It extended past the bronchi – the main respiratory pathways into the lungs – right down into the air sacs or alveoli that make up their very substance.

All the viruses that Taubenberger's group had sequenced so far came from individuals who had died in the autumn of 1918 – during the most deadly wave of the pandemic.[5] But the AFIP repository also contained tissue from spring-wave victims. In 2011 – by which time he had moved to the Laboratory of Infectious Diseases at the National Institutes of Health (NIH) in Bethesda, Maryland – Taubenberger published a comparison of the gene sequences encoding the H antigen from the two waves. From this it became clear that the virus had undergone a small but critical change between the spring and the autumn, such that the H antigen was now less well adapted to birds, and better adapted to humans. Three-quarters of the spring-wave cases had a bird-adapted H, while three-quarters of the autumn cases had a human-adapted one.

Since the vast majority of those who caught Spanish flu recovered, focusing on the few who died risks distorting the picture. However, the NIH team has also studied medical records that were kept in US military camps between 1917 and 1919, that logged both lethal cases and cases where the patient recovered. They show that, while the overall number of cases of influenza dwindled between April and August 1918 – that is, between the spring and the autumn waves – the proportion of them that were complicated by pneumonia rose steadily over the same period. The lesions that the flu virus creates in the lining of the respiratory tract can become infected by bacteria, resulting in pneumonia. In Taubenberger's view, the worse the underlying flu, the more likely it is to invite in opportunistic bacteria. Hence he regards the 1918 label of pneumonia as a marker flagging up the presence of the highly virulent, pandemic virus. If he's right, then over the summer of 1918, that virus acquired the capacity to spread easily between humans.[6]

Bringing all the evidence together, Taubenberger now believes that the virus emerged through a background of seasonal flu sometime in the winter of 1917–18, and was already circulating at low levels the following spring. Whether it came directly from a bird, or passed via a pig, he can't yet say. In the summer of 1918, it mutated, becoming highly contagious between humans. This new, more virulent form spread through the viral population that summer, and in the autumn the disease erupted. By then, the seasonal background had receded, and there was nothing to dilute the 'pure' pandemic variety.

What caused the virus to mutate that summer is not clear, but as we've seen, flu doesn't need much prompting to change, and conditions were arguably conducive to such an event. Large parts of the world were in the grip of famine, and there is some evidence that nutritional deficiencies in the host can drive genetic changes in the flu virus, causing it to become more virulent (while simultaneously impairing the host's immune response).[7] If we accept that the second wave emerged on or close to the Western Front, then that front was awash with chemicals, some of which, mustard gas in particular, were mutagenic – meaning they were capable of inducing genetic changes in living organisms, including viruses. And those same gases had compromised the lungs of many of the young men gathered there, rendering them ripe for invasion.

Evolutionary biologist Paul Ewald has even argued that the ratcheting up in virulence of the flu virus that summer was a direct response to conditions on the Western Front.[8] It is often said that the optimal strategy for an agent of infectious disease that is transmitted directly from host to host is to moderate its virulence, so that an infected host remains alive for long enough to spread the disease far and wide. But if the pool of hosts is not very mobile – its movement being limited by being packed into trenches, say – and if those hosts are dropping dead from other causes, then there is less evolutionary pressure on the virus to moderate its virulence. In those conditions, Ewald says, there is no advantage to it keeping its host alive. Of course, the virus has no strategy in the conscious sense of the word. Rather, highly virulent strains come to dominate the viral population through natural selection, because they are the most likely to survive and reproduce.

The human immune system takes several years to mature, and in old age it loses its potency. This is the explanation that is usually given for flu's characteristic U-shaped death curve. But in 1918 adults in the prime of life also died in large numbers. Some have suggested that it was precisely because their immune systems were so robust that they were vulnerable, since it was in them that the cytokine storm was most aggressive. There is a problem with that explanation, however. As far as we know, the immune system is just as robust in a fifteen-year-old as it is in a twenty-eight-year-old, yet in 1918, fifteen-year-olds were down there in the first trough of the W: though they got ill in large numbers, relatively few of them died. And something else needs explaining: the W was not symmetrical. The right-hand upstroke was attenuated, meaning that the aged were in general *more* protected than usual. They were actually less likely to die in the 1918 pandemic than they had been in seasonal flu outbreaks throughout the previous decade.

The answers to these puzzles may lie in the different age cohorts' previous exposure to flu. There is a school of thought that holds that the immune system's most effective response to flu is to the first version of the virus it ever encounters. All subsequent exposures elicit variations on that response that are never a perfect match for the new strain. There are hints, based on tests of the antibodies present in blood taken from people who were alive in the first half of the twentieth century, and stored ever since, that the flu subtype responsible for the Russian flu of the 1890s was H3N8. If so, then those who were aged between twenty and forty in 1918, for whom the Russian flu was probably their first exposure to influenza, were primed to deal with a very different subtype to the Spanish flu, and consequently produced an inadequate immune response in 1918. By the same logic (though there are as yet no serological data to support this hypothesis), the very old may have been afforded some protection in 1918, by virtue of having been exposed to a flu subtype containing either H1 or N1, that circulated in humans around 1830.

What about the question of where the Spanish flu came from? We would like to know the answer to this, because it might help us to identify the conditions that give rise to a so-called 'spillover' event – when a virus 'jumps' the species barrier – and reduce the chances, as far as possible, of it happening again. In order to choose between the three current theories, or indeed, to identify a geographical origin that no one has yet proposed, scientists would need to compare the sequence of the virus that caused the Spanish flu with those of viruses that caused earlier outbreaks of respiratory disease in those places. They can't do that yet, because the oldest human flu sequences on record belong to the Spanish flu itself. Given that, to date, they have found that virus almost everywhere they have looked for it – eventually, and with the help of intrepid flu hunters like Johan Hultin – it is possible that viable samples will still come to light that will enable them to make those comparisons. That would be the holy grail for Jeffery Taubenberger. In the meantime, however, they haven't been idle. Other researchers have been using a new technique to make educated guesses about which of the proposed origins is most likely.

The technique in question is based on the concept of a 'molecular clock'. Every living organism must copy its genetic material in order to reproduce, but as we've seen, the mechanism by which it does so is not perfect, and flu's copying mechanism is particularly error-prone. Some errors shape the virus – we call them, cumulatively, drift – but the majority are 'silent', meaning they have no effect on its structure or function. In any given host, these silent errors build up at a constant rate, which means that by counting the genetic differences between two related viruses, you can obtain a measure of the time that has elapsed since they split from a common ancestor. This is a molecular clock: it has nothing in common with a real clock, except that it counts time.

Flu infects many animals – not just humans, birds and pigs, but also dogs, horses, bats, whales and seals. At the University of Arizona, evolutionary biologist Michael Worobey has compared all the available sequences of flu viruses that are currently circulating or have circulated in different hosts over the last century, and used them to build a family tree of influenza. The virus accumulates errors at different rates in different hosts, but because he knows that, and has calculated those rates, he can make retrospective predictions about when various historical strains were born, and with what parentage. In 2014, Worobey reported that seven of the eight genes in the 1918 virus closely resembled flu genes found in birds in the western hemisphere – in North America, to be precise.[9]

Does that lay to rest all the fevered speculation about the origins of the Spanish flu? Did it begin in Kansas after all? Worobey's work is suggestive, without being definitive. Molecular clocks are, in general, not as reliable as comparing actual sequences. Nevertheless, they have been right before. In 1963, flu broke out in horses in racing stables in Miami, eventually spreading to horses throughout the United States. Worobey found that the horse flu strain was related to one then circulating in birds in South America, corroborating contemporary veterinarians' reports that the flu had probably reached Miami with some thoroughbreds that had been flown in from Argentina.

Questions remain, not least over that troublesome eighth gene – the one that encodes the H1 antigen – which seems to tell a different story. The flu family tree indicates that it may have been circulating in humans for a decade or more prior to 1918, at which point it recombined with seven bird flu genes in a shift event that produced the Spanish flu. If that is what happened, it could explain that troubling cohort of the five-to-fifteen-year-olds, who got sick in droves but didn't die, since they would have been exposed to the H1 antigen as babies, and been forearmed against it. That scenario raises questions of its own, however, not least why the emergence of that antigen in humans didn't trigger a pandemic earlier. While scientists continue to scratch their heads over that problem, the molecular clocks have a few more insights to offer, and these may be the most troubling yet.

The current consensus is that, for hundreds of thousands if not millions of years, wild birds have harboured a kind of primordial soup of flu viruses, some of which occasionally infect humans. The assumption is that – as with HIV, which came from monkeys inhabiting African forests – we unwittingly disturbed a pre-existing reservoir, allowing the virus to move into humans. But things may have happened differently, and we may be a much more central player in the flu ecosystem than we think.

While teasing out the flu family tree, Worobey noticed that much of the branching in bird flu lineages was relatively recent, meaning that birds are a young reservoir in evolutionary terms. As recently as biblical times, or 212 BC, when flu ravaged the armies of Rome and Syracuse in Sicily, humans might have been more likely to catch flu from horses – another animal with which people have lived closely since the farming revolution. At some point in the last 2,000 years, birds took over as the more important reservoir. It looks as if the bird flu lineage that contributed most of its genes to the 1918 human strain became established in North America around the same time as an epidemic of horse flu that broke out in Toronto in 1872 and spread throughout that continent (newspapers described the almost deserted streets of Washington DC, and a backlog of freight at rail terminals in Philadelphia, as sick mules and horses were taken out of circulation). Worobey can't yet tell if that flu passed from horses to birds, or vice versa, but one could speculate that the switch occurred as a result of horses being replaced by mechanised modes of transport, and the expansion of poultry farming in the late nineteenth and early twentieth centuries. The switch had occurred by 1918, but its legacy was that horses – like humans – were now vulnerable to infection by bird flu. In fact, army vets in several warring nations – as well as in the neutral but conflict-ready Netherlands – reported an epidemic of horse flu in cavalry stables, that coincided with the human one.[10]

The suggestion is that we humans actively drew animal reservoirs of flu into our midst – and even created new ones – through our domestication of wild animals. If so, then the greatest threat to our health, in terms of the next flu pandemic, may not be wild birds. It may be much closer to home. Ducks are not the only avian incubators of flu, but as Claude Hannoun and others discovered in the 1970s, they are particularly effective ones. Archaeological evidence suggests that they were first domesticated in southern China around 4,000 years ago. Today there are an estimated 1 billion domestic ducks in the world, meaning they probably already outnumber wild ducks, and there is no ecological barrier between the two. The Chinese, for example, herd ducks through rice paddies to eat insects and other pests, and there they mingle with wild birds. For at least 150 years – that is, since before the Spanish flu – flu genes have been as likely to flow from domesticated birds to wild birds as in the opposite direction. Thanks to our animal-husbandry practices, in other words, we now pump flu genes into nature. The 1918 flu virus may have jumped to humans from a wild bird (either directly or via a pig), but it is just as likely to have come from one raised in a farmyard.

Our right to blame the other is looking distinctly shaky. If the molecular clocks are right, humans contributed to their own misery both in 1918 and since. There were two further flu pandemics in the twentieth century: the 1957 'Asian' flu, which claimed 2 million lives, and the 1968 Hong Kong flu, which killed perhaps twice that. They were caused by subtypes H2N2 and H3N2 respectively, but both inherited the lion's share of their internal genes from the 1918 flu, causing Taubenberger and his colleague, epidemiologist David Morens, to dub the Spanish flu 'the mother of all pandemics'.[11] In the 1930s, the British and American teams who demonstrated that flu was caused by a virus surprised their peers by claiming that humans may have passed the Spanish flu virus to pigs, and not vice versa. Comparisons of human and pig flu sequences have since confirmed their suspicions, and in 2009 the H1N1 subtype that had been circulating in pigs since 1918 erupted again in humans, in modified form, triggering the first flu pandemic of the twenty-first century. It was dubbed 'swine flu', for obvious reasons, though in a longer timeframe it was humans who gave it to humans. Swine were mere intermediaries.

15

The human factor

There was still one very large, unexplained puzzle. Granted, twentysomethings were vulnerable, but why were some twenty-somethings more vulnerable than others? Why did the impact of the Spanish flu vary over *space* as well as *time*, such that in a given age group, more Kenyans died than Scots, more Indonesians than Dutch? In a future pandemic, would *you* die? Would your sister living on another continent? Which of your children would be more likely to survive? If we knew who was vulnerable, we could take steps to protect them.

To understand what it was that caused some people to succumb and others to get off lightly, we have to follow the numbers. In 1918, people were struck by the terrifying randomness with which the flu seemed to choose its victims. It was only when scientists started comparing morbidity and mortality rates that they began to discern certain patterns. This led them to conclude that humans themselves had shaped the pandemic – through their unequal positions in society, the places where they built their homes, their diet, their rituals, even their DNA.

First, let's sketch out that geographical unevenness with a whistlestop tour of the world in figures – excess mortality rates, to be precise. These varied to an astonishing degree. If you lived in certain parts of Asia, you were *thirty times* more likely to die from flu than if you lived in certain parts of Europe. Asia and Africa suffered the highest death rates, in general, and Europe, North America and Australia the lowest. But there was great variation within continents too. Denmark lost approximately 0.4 per cent of its population, Hungary and Spain around three times that. African countries south of the Sahara experienced death rates two or even three times higher than those north of the desert, while in Asia the rates varied from roughly 2 per cent in the Philippines, to between 8 and 22 per cent in Persia (the large range reflects the fact that Persia was in crisis, and gathering statistics was hardly anybody's priority). India, which included Pakistan and Bangladesh at the time, and which lost around 6 per cent of its population, suffered the greatest loss in absolute numbers of any country in the world. Between 13 million and 18 million Indians died, meaning that more Indians may have died of Spanish flu than human beings were killed in the First World War.

Cities tended to suffer worse than rural areas, but within a country, some cities fared worse than others. Thus Chicago got off lightly compared to Washington DC, which got off lightly compared to San Francisco. Within cities, there was variation. In the Norwegian capital Kristiania (Oslo), for example, death rates rose as apartment sizes shrank.[1] In Rio, it was the mushrooming *subúrbios* – sprawling shanty towns at the edge of the city – that suffered the heaviest losses. Newly arrived immigrants tended to die more frequently than older, better-established groups – though the pattern is sometimes hard to discern, because there are fewer data for the immigrants. A 1920 study of what happened in the US state of Connecticut nevertheless reported that 'The Italian race stock contributed nearly double its normal proportion to the state death roll during the epidemic period.' The Italians, as we know, were the newest immigrant group to arrive in America. In fact, residents of Connecticut who were of Italian origin were more likely to die than those of Irish, English, Canadian, German, Russian, Austrian or Polish background.[2]

What caused these inegalities? Some of it comes down to disparities in wealth and caste, and – as far as it reflected these – skin colour.[3] Eugenicists pointed to the constitutional inferiority of the 'degenerate' races, whose lack of drive caused them to gravitate to squalid tenements and *favelas*, where the diseases to which they were prone naturally followed them (in other words, they argued that Italians were more susceptible because they were Italian). In fact, it was bad diet, crowded living conditions and poor access to healthcare that weakened the constitution, rendering the poor, immigrants and ethnic minorities more susceptible to disease. This was why, in Korea, ethnic Koreans and Japanese people fell ill at roughly similar rates, but Koreans were twice as likely to die.[4] And why, in India, the remote, forested region of the Dangs in Gujarat lost a higher proportion of its population than most Indian cities (16.5 per cent between 1911 and 1921, mainly due to the Spanish flu). The Dangs bucked the 'rural advantage' trend, probably because they were home to *adivasis* – the so-called original inhabitants of the area – who were looked down upon by both the British and other Indians as backward jungle tribes.[5]

Statisticians were foxed by their observation that the highest death rates in the French capital were recorded in the wealthiest neighbourhoods, until they realised who was dying there. The ones coughing behind the grand Haussmannian facades weren't the owners on the *étage noble*, but the servants in the *chambres de bonne*. As Theresa McBride explained in her book, *The Domestic Revolution*, 'Close enough to their employers' apartments on the floors below, the servants were segregated into a society of their own where they need not be seen but could be easily summoned.' They worked fifteen-to-eighteen-hour days and often had to share their sleeping spaces with other servants. 'The servant's room was generally small, with sloping ceilings, dark, poorly ventilated, unheated, dirty, lacking privacy or even safety,' wrote McBride. The flu may have been democratic, as one French historian pointed out, but the society it struck was not: a quarter of all the women who died in Paris were maids.[6]

There were other paradoxes. African Americans, though severely discriminated against in the United States, seem to have had a light dose – and they noticed it at the time. 'As far as the "Flu" is concerned the whites have the whole big show to themselves,' wrote one J. Franklin Johnson to the *Baltimore*

Afro-American, adding that, had it been otherwise, 'we would have never heard the last of it, and health talks to colored people would have been printed by the wholesale in seventy-two-point type in the daily papers'.[7] The case of the African Americans remains puzzling today (were they disproportionately exposed to the mild spring wave, and so protected, to some extent, against the autumn one?), but another mystery has been solved: the discrepancy in the death rates at the Rand gold mines and the Kimberley diamond mines in South Africa. This, it seems, came down to those black tentacles of rail tracks.[8]

The gold mines were by far the bigger of the two operations, employing almost twenty times as many men as the diamond mines, and Johannesburg – the city built to serve them – was consequently the larger rail hub. The rail network connected Johannesburg to the country's east coast, and in particular to Durban, the major port of Natal Province. Though South Africa didn't record a 'herald' wave of Spanish flu, as such, an epidemiologist named Dennis Shanks has found reports buried in the literature of cases of a mild, flu-like illness that arrived on ships in Durban in July 1918. From there, the infection travelled northwards, along the rail tracks, to the Rand. When the flu returned to the Rand a few months later, therefore, the gold miners may have been partially protected. Kimberley, on the other hand, was relatively poorly served by the rail network. Lying 500 kilometres to the south-west of Johannesburg, it was however connected to Cape Town, and it therefore received its first dose of flu from that city after the arrival of the infected troopships *Jaroslav* and *Veronej*, having had no prior exposure. When the flu left the industrial centres with the panicked miners, Natal Province was again protected compared to those parts of the country that were not served by the Durban–Johannesburg branch of the rail network – notably the Transkei and Ciskei regions of the Cape, where death rates were three times higher than in Natal.

Isolation was also the reason that some of the most remote places on earth were vulnerable. Lack of historical exposure to the virus translated into higher death rates, which were often amplified by problems associated with poverty and exclusion. After the steamer the SS *Talune* left Auckland, New Zealand with infection on board, it carried that infection to a string of Pacific Islands in turn. Fiji subsequently lost around 5 per cent of its population, Tonga twice that, and Western Samoa a staggering 22 per cent.

Cities were more vulnerable to infection than rural areas mainly because of the density of their populations, but what about that puzzling variation *between* cities? Exposure to the mild spring wave may have buffered those that received it, but an effective disease-containment strategy also had an impact. One 2007 study showed that public health measures such as banning mass gatherings and imposing the wearing of masks collectively cut the death toll in some American cities by up to 50 per cent (the US was much better at imposing such measures than Europe). The timing of the measures was critical, however. They had to be introduced early, and kept in place until after the danger had passed. If they were lifted too soon, the virus was presented with a fresh supply of immunologically naive hosts, and the city experienced a second peak of death.[9]

In Zamora, mass gatherings were positively encouraged – and at 3 per cent, or more than twice the national average, Zamora had the highest death rate of any city in Spain. In fact, religious rituals – or secular rituals masquerading as religious ones – contributed to the shape and possibly the duration of the pandemic everywhere. Some have argued, for example, that there were really only two waves of sickness – in the (northern hemisphere) spring and autumn of 1918 – and that what appeared to be a third wave in early 1919 was simply the tail end of the second after a brief hiatus due to end-of-year festivities. Around the time of Christmas and Hanukkah, for example, Christian and Jewish children tended to stay away from school, depriving the virus of a valuable pool of potential hosts – until, that is, they returned to class in the New Year.

An underlying disease made you more susceptible to the Spanish flu. Medical historian Amir Afkhami has suggested that Persians fighting in the British Army were more severely affected by the flu than native British soldiers, because they were more likely to be suffering from malaria and its complication anaemia (a reduction in the number of red blood cells or haemoglobin – the oxygen-carrying molecule – in blood), which impairs the immune response.[10] The pandemic also purged the world of a disproportionately large number of TB patients who would otherwise have died more slowly over the following decade. In fact, it is possible that TB – once described as 'the captain of all these men of death', because of the misery it inflicted throughout the nineteenth century and into the twentieth – was one of the main reasons why the flu killed more men than women globally. In that vulnerable twenty-to-forty-year age group, more men than women had TB, in part because they were more likely to be exposed to it in the workplace.[11]

Thus culture shaped biology: men were more likely to go out to work, in many countries, and women were more likely to stay at home. But although more men than women died overall, in some countries that trend was reversed in certain age groups. In India, strikingly, it was reversed in *every* age group. Why did more women die than men in India, when Indian women were also traditionally home-makers? One argument goes like this: in times of crisis, Indian girls and women – who were already more likely to be neglected and underfed than their male counterparts – were also expected to care for the sick. They therefore had both greater exposure and less resistance to the disease, and dietary taboos may have exacerbated their susceptibility.

The main religion in India is, and was, Hinduism. Hindus are not necessarily vegetarian, but a vegetarian diet is associated with spiritual serenity, women are more likely to be vegetarian than men, and traditionally, vegetarianism is obligatory for widows. In her detailed analysis of life in a northern Indian village in the 1920s, American anthropologist and missionary Charlotte Viall Wiser noted that the villagers' diet consisted predominantly of what their fields could furnish – cereals, pulses and vegetables. She was astonished to find that most of them did not lack iron (iron deficiency is a common cause of anaemia), but she described how they eked every last atom of it from their food. Grains, for example, were not milled but eaten whole, with the iron-rich outer layers intact. She felt that they lived at the margin of deficiency, and that any slight disruption could push them over the edge.[12] The drought that followed the failure of the south-west monsoon in the summer of 1918 certainly qualified as such a disruption.

When all other things were equal, when neither wealth nor diet nor festival season nor travelling habits could differentiate two groups of human beings, there was still that distressing residual disparity that meant that one might be decimated while the other survived more or less intact – as if a god had thrown his thunderbolts carelessly. Death struck very unevenly across the territory of Alaska, for example. Bristol Bay, the worst-affected region, lost close to 40 per cent of its population, but other parts lost less than 1 per cent – on a par with some of the great American metropolises – and a relatively high number of Alaskans, one in five, escaped the disease entirely. This was the recalcitrant core of the variation, and for a long time it defied explanation. Many wondered if the answer lay in human genes – in the way they shaped the host-virus encounter – but how to prove it? People who share genes often share an environment too, which is another way of saying that families tend to live together, so they are exposed to the same germs. Disentangling the two effects would not be easy.

The Mormons inadvertently provided a way to slice the knot. Mormons, members of the Church of Jesus Christ of Latter-day Saints, believe that the family unit can survive death if all of its members have been baptised, and those who weren't baptised in life may be baptised after death. They are therefore conscientious genealogists who keep detailed records of their family trees, which they store on millions of rolls of microfilm in a vault under

Granite Mountain – a peak in the Wasatch range close to Salt Lake City. The vault, which was built in 1965, is protected by a thirteen-tonne steel door designed to withstand a nuclear explosion, but these days you can access the archives via the Internet. Even more helpfully, the records have been digitally linked to the relevant death certificates, meaning that it is possible to learn at the tap of a key what an individual Mormon died of. In 2008, Frederick Albright and colleagues at the University of Utah identified close to 5,000 Mormons who had died of flu in the previous hundred years. Having reconstructed their family trees, they discovered that a blood relative of one of these index cases was more likely to have died of flu than an unrelated person, even if the two relatives had never shared an environment.[13]

It was a fascinating hint that flu might have a heritable component, but other studies failed to replicate the finding. Then in January 2011, in the midst of the annual flu season in France, a two-year-old girl was admitted to the intensive care unit of the Necker Hospital for Sick Children in Paris, suffering from ARDS (acute respiratory distress syndrome). Doctors saved her life, and one of them, Jean-Laurent Casanova, sequenced her genome. He wanted to know if it held the key to why an otherwise healthy child had nearly died of a disease that most children shrug off. It turned out that the girl had inherited a genetic defect that meant she was unable to produce interferon, that all-important first-line defence against viruses. As a result, her besieged immune system went straight to plan B: a massive inflammatory response similar to the one pathologists saw in 1918. The child's genetic defect was rare, but Casanova went on to identify a cluster of similar defects that also result in an inability to make interferon. Collectively, he calculated, these might affect one in 10,000 people – roughly the incidence of ARDS during an annual flu outbreak.[14]

What Casanova's finding meant was that, regardless of their culture, diet, social status or income, one in 10,000 people are particularly vulnerable to flu – a vulnerability that they inherit from their parents. In the 1918 pandemic, these people were probably among the first to die, but a hundred years on, we have it in our power to level the genetic playing field and give them a fighting chance. The reason is that the genetic defect that prevents an individual from making interferon does not affect his or her ability to produce antibodies. In theory, therefore, such a person can be protected against flu just by being vaccinated with the standard annual flu vaccine. Every year since 2011, the girl whom Casanova first met in the ICU of the Necker children's hospital has received a shot of flu vaccine and sailed through the subsequent flu season as easily as her peers.

Casanova had discovered a genetic component to flu, and perhaps, the last piece in the puzzle of why the Spanish flu struck so unevenly. His finding fell on fertile ground, because at the time, scientists were beginning to think about infectious diseases in a new way – that is, as partly genetic. The idea they were working towards was this: *all* infectious diseases have a genetic component, but in some, one or a few genes control susceptibility to that disease, while in others the genetic component consists of the small, cumulative effects of many genes. In the first case, a defect in one of those genes causes a large increase in susceptibility; in the second, only a small one. If this idea turns out to be correct, we will have to recalibrate the way we think about disease yet again: not only might infectious diseases be partly genetic, but diseases that we have long thought of as genetic or 'environmental' in origin might turn out to be partly infectious, too. One theory about Alzheimer's disease holds, for example, that it is caused by 'prions' – infectious agents that, until recently, were as shrouded in mystery as viruses were in 1918.

A hundred and fifty years ago, George Sand was affronted when the residents of Palma, Majorca asked her to leave on the grounds that her lover's disease was infectious rather than heritable. Today, we know that TB is caused by the bacterium *Mycobacterium tuberculosis*, but that susceptibility to that bacterium is inherited. Something similar applies to influenza – a disease that, a hundred years ago, was thought to be bacterial. To the best of our knowledge in 2017, flu is caused by a virus, but it is also partly under the control of human genes. Understanding that helps us to make sense of the extraordinary variability in its manifestation, that people found so baffling in 1918. They couldn't see beyond the surface phenomena; now we're able to look 'beneath the bonnet'. (One day, science might help us to explain diseases that mystify us today for the same reason, such as autism spectrum disorder.)

The revision in how we think about flu seems radical, but perhaps it isn't as radical as all that. While observing sick silkworms in the nineteenth century, Louis Pasteur made two observations: first, that *la flacherie*, as the worms' disease was called (literally, 'flaccidity' – caused by eating contaminated mulberry leaves, it gave them debilitating diarrhoea) was infectious; and second, that offspring could inherit it from their parents. In all the furore over the first observation, the second was overlooked. Perhaps the time for Pasteur's second insight has finally come.

PART SEVEN: The Post-Flu World

Linus H. French with some of the flu orphans rescued from Bristol Bay, Alaska, 1919

16

The green shoots of recovery

In February 1919, Adam Ebey and his wife Alice took a train into the hills of Gujarat, then trekked forty kilometres through teak and bamboo jungle to the Church of the Brethren mission at Ahwa, which they were due to take over. The third wave of the pandemic had just erupted, many of the *bhagats* or traditional healers had fled, and the services of the 'surgeon sahib', as Ebey would become known, were immediately in demand. Once the sickness had passed, however, he sat down to write to the home church in Illinois, and recounted the story of Laksman Haipat.

Haipat was a twenty-five-year-old farmer and Christian convert who, having already been widowed once, had married for a second time in January 1919. Not long after the second wedding, he left his village on business. When he returned a few days later, he found it deserted and his new bride lying under a tree, in the terminal phase of Spanish flu. He stayed with her until she died, then he dug her grave. 'She was a heavy woman,' Ebey wrote. 'He could not carry her, so he took a rope and dragged her into the grave. What else was there for him to do? He married his third wife the day after Christmas, 1919.'[1]

Before the pandemic, mortality rates had been in decline the world over, partly due to advances driven by germ theory; the pandemic reversed that trend for three years. India paid a particularly high price – so high, in fact, that in 1964 the Nobel Prize-winning economist Theodore Schultz used what happened in that country to test a theory that there is surplus labour in traditional farming systems. He concluded that there isn't, since India's post-flu agricultural output shrunk by 3 per cent, compared to pre-1918 levels. But humans are supremely resilient, and the recovery seems to have begun almost immediately the shock had passed. Though India saw a 30 per cent reduction in births in 1919, starting in 1920, fertility not only returned to pre-flu levels, it exceeded them – triggering what has been described as the beginning of a demographic revolution.[2]

It wasn't only India that saw a rebound. Fertility rates collapsed in Europe around 1918, only to recover spectacularly two years later – briefly attaining levels higher than those seen prior to 1914. Most observers put this down to the war, and to a wave of conception that followed the men's return. But that doesn't explain why neutral Norway also saw a baby boom in 1920. Norwegian men did not go away to fight, but like the rest of the Norwegian population, they did suffer from flu. 15,000 Norwegians died in the pandemic, and there were 4,000 fewer conceptions in 1918 than would have been expected, but the following year saw conceptions surpass what was needed to make up for that shortfall by 50 per cent. In other words, three babies were conceived in 1919 for every two that had not been in 1918.[3] Could the flu have contributed to a global baby boom? Indeed it could, and the explanation lies in the way it chose its victims.

The pandemic took time to recede: the middle peak of the W-shaped mortality curve shrank until, between 1922 and 1928, depending on where you were in the world, the W bottomed out in a U.[4] The Spanish flu had been tamed, but it had left behind a very changed humanity. By purging the less-than-fit – those who were already sick with malaria, TB and other diseases – it had created a smaller, healthier population that was now able to reproduce at higher rates. This is one theory for why fertility rebounded so dramatically: survivors like Laksman Haipat now married other survivors who were, by definition, healthier and more robust than those who had perished.

Can we really say that humans were healthier in the wake of the Spanish flu? It seems extraordinary, but in one very crude way, we can: their biological capacity to reproduce increased, and they had more children. It's crude, because other factors besides biology shape the number of children people have – religious and economic considerations, for example. But there are other indications that men, in particular, were healthier – namely, that their life expectancy increased. Before 1918, women had lived on average close to six years longer than men. The flu killed roughly 170 more men than women per 100,000 people, and by the time the pandemic was over the gap in life expectancy had closed to one year. Women wouldn't regain their previous advantage until the 1930s, mainly because, by then, heart disease had become both much more prevalent, and more male.[5]

Overall, therefore, it can reasonably be argued that the post-flu population was healthier. If we look more closely, however, we see a more nuanced picture, in which some groups were certainly worse off than before. First, let's consider those who were in their mother's womb in the latter half of 1918. Pregnant women, as we've said, were extremely vulnerable to the Spanish flu – and this was true the world over. They were 50 per cent more likely than women who weren't pregnant to develop pneumonia, according to one 1919 estimate, and 50 per cent more likely to die if they did.[6] It's not clear why, but one suggestion is that the culprit was not the virus itself, but the cytokine storm it provoked – that deluge of chemical alarm signals that diverted blood and immune cells to the lungs. Already subject to the physiological stresses of carrying a fetus, pregnant women might have succumbed to this more easily, and if blood was diverted away from the womb, that might also explain why miscarriages were so common. The dent in the population left by the loss of those unborn children is only just now working itself out, as we pass their projected lifespan. But some of those children *were* born, so the question arises: what mark, if any, did the flu leave on them?

A baby born in 1919, who had weathered the slings and arrows of the Spanish flu inside his mother's womb, and who turned up at a military recruitment depot in 1941, was an imperceptible 1.3 millimetres shorter, on average, than recruits who had not been exposed to it prenatally. That may not seem like much, but it is an indication that the stresses affected every fetal organ, including the brain. As his life unfurled, this child was less likely to graduate and earn a reasonable wage, and more likely to go to prison, claim disability benefit, and suffer from heart disease after the age of sixty.[7]

Only men took the draft in 1941, but the same applied to anyone unlucky enough to be conceived in the first months of 1918, regardless of their sex or colour: they were a diminished generation. The British writer Vera Brittain – who nursed at Étaples during the First World War – applied the term 'lost generation' to those well-born, educated young men who died while serving with the British Army, and who might have gone on to great things had they returned. But those who were in their mother's womb when the Spanish flu struck, who are often held up as an example of why it's important to invest in the health of pregnant women, were the twentieth century's real lost generation.

Others were left worse off too. There is good evidence, for example, that the Spanish flu was itself a chronic disease, and that it had a negative impact on some people's health for months or even years after the initial flu-like symptoms had subsided. Hungarian composer Béla Bartók was left with a severe ear infection that made him fear permanent deafness – ironically, the fate of his musical hero, Beethoven. He took opiates for the pain, but they couldn't banish the auditory hallucinations that plagued him for some time after. For American aviator Amelia Earhart, meanwhile, the legacy was a lifelong sinusitis that, some say, affected her balance and ability to fly. The first woman to fly over the Atlantic, in 1928, she vanished while flying over the Pacific nine years later.

Previously we saw that feelings of anxiety accompanied the acute phase of the disease, and that there were instances of people killing themselves while delirious. If and when they recovered from that phase, however, some patients found themselves plunged into a lingering state of lassitude and despair. How much of this wave of 'melancholia' was due to the flu, and how much to the war? It's a difficult question. The flu virus may act on the brain, causing depression, but depression is also a common response to bereavement and social upheaval. How to disentangle the two? Once again, a study from neutral Norway may help.

Norwegian epidemiologist Svenn-Erik Mamelund studied asylum records in his country from 1872 to 1929 and found that, in every year in which there was no pandemic of influenza, a few cases were admitted of mental illness associated with flu. In each of the six years following the 1918 pandemic, however, the average number of such admissions was *seven times* higher than in those non-pandemic years. Because it's hard to know exactly what those patients were suffering from, and impossible to retrospectively demonstrate a link between their psychiatric symptoms and flu, any conclusions drawn from these data must be tentative. With that caveat, however, Mamelund speculates that the patients admitted in those six years were survivors of Spanish flu who were suffering from what today we would call post-viral or chronic fatigue syndrome. He also believes that they were the tip of the iceberg, since in those days, most people suffering from melancholia would not have sought out a psychiatrist.

Intriguingly, one Norwegian seems to have escaped the melancholia this time around: Edvard Munch. One might consider him unlucky to have been caught up in two flu pandemics, but we can't be sure that he was a victim of the Russian flu, and the idea that it influenced his painting of *The Scream* is therefore pure speculation. On the other hand, he almost certainly did suffer from the Spanish flu, and as he recovered from it he painted a series of self-portraits, one of which shows him sitting, yellow and gaunt, in a wicker chair. Some have suggested that these paintings depict his post-viral melancholia, but his biographer Sue Prideaux disagrees. He was melancholic by nature, she says, but after the flu he entered a highly creative period. He painted at least fourteen important works in 1919, and they are striking for their optimism and celebration of nature. 'Their colours are clear, the hand is steady, the vision and power is undiminished,' writes Prideaux.[8]

We don't know how many people suffered from depression in the wake of the Spanish flu, but the Norwegian wave is unlikely to have been unique. Post-viral syndrome has been blamed, for example, for triggering the worst famine in a century in Tanzania, where crippling lethargy prevented an already depleted population from planting when the rains came at the end of 1918. The 'famine of corms', as it was known (the name refers to the root structure of the banana plant, that African women fed their families in times of hunger), lasted for two years.

Often the psychiatric symptoms were temporary. In 1919, for example, 200 'recovered' flu patients were admitted to Boston Psychopathic Hospital suffering from delusions and hallucinations. Around a third of them were diagnosed as having dementia praecox, an obsolete name for schizophrenia. Dementia praecox was supposed to be incurable, but five years later most of them had made a full recovery. The psychiatrist who followed the Boston patients, Karl Menninger, thought a new diagnostic label was called for, to describe this acute, reversible schizophrenic syndrome that had come in the wake of flu.[9]

One other neurological condition has been associated with the Spanish flu, and that is encephalitis lethargica (EL), or more colloquially, 'sleepy sickness'. EL washed over the world in a wave between 1917 and 1925, peaking in 1921. It came on with flu-like symptoms and, as its name suggests, overwhelming sleepiness. But it was a strange sleepiness: though patients showed all the outward signs of somnolence, they seemed to remain aware of their surroundings. One female patient, filmed in Germany in 1925, fell asleep while moving her finger to her nose in a test of coordination – but continued, if somewhat erratically, to complete the task.[10] A third of those who contracted EL – an estimated half a million people worldwide – died within a few weeks. Another third recovered, while the remainder went on, after a delay that could stretch to years, to develop a form of paralysis that resembled advanced Parkinson's disease.

Was the epidemic of EL related to that of the Spanish flu? The question has been debated since the 1920s. Those who believe that it was point to the following 'smoking guns': although cases of EL have been reported at other times in history, the surge in the 1920s is the only recorded epidemic; some of the earliest cases in that epidemic were recorded on the Western Front in the winter of 1916, around the time of the outbreak of purulent bronchitis; Western Samoa, which suffered so badly from the flu, experienced a wave of EL, but American Samoa escaped both; and the average age of those affected was twenty-nine.

Smoking guns are one thing, hard evidence is another, and to date scientists have failed to establish a causal link between the two epidemics. They know that the flu virus can travel up the olfactory nerve from the nose to the brain, causing inflammation there, and potentially triggering seizures and strokes (encephalitis is another name for brain inflammation), and they acknowledge that EL behaved suspiciously like a viral disease in its early stages. But they have yet to find viral RNA in brain tissue taken from EL patients post-mortem. That doesn't mean it's not there – it may simply be that their techniques aren't yet sensitive enough to detect it – so for the time being, the jury is out.[11]

None of those flu patients who went on to suffer long-term neurological or psychiatric conditions were 'purged' from the population, in the literal sense of dying, but societies often found other ways to exclude them. Nontetha Nkwenkwe, for example, was locked up for recounting her fever-fuelled dreams. We'll tell her story in the next section, but we'll end this one with the eloquent case of Rolando P. Mr P was one of the unfortunate third of EL patients whom the disease left imprisoned in their own bodies, and who usually ended up institutionalised and forgotten, their faces unlined even in old age. The British neurologist Oliver Sacks told the story of a group of such patients – and their temporary reprieve thanks to the Parkinson's drug L-dopa – in his bestselling book *Awakenings* (1973). Rolando P was one of them.

Born in New York in 1917 to a newly immigrated and very musical Italian family, Mr P contracted a fever at the age of three. This illness, which was accompanied by intense drowsiness, lasted for more than four months, and when he 'woke' from it, his parents realised that he had undergone a profound change. His face was expressionless, and he could barely move or speak. For a few years he attended a school for the mentally defective, but his lack of balance made school increasingly problematic, and eventually his parents stopped sending him. 'From his eleventh to his nineteenth year, he remained at home, propped before the speaker of a large Victrola gramophone, for music (as his father observed) seemed to be the only thing he enjoyed, and the only thing which "brought him to life".' In 1935 Mr P was admitted to the Mount Carmel Hospital in New York, and as Sacks put it, 'The next third of a century, in a back ward of the hospital, was completely eventless in the most literal sense of this word.'[12]

NONTETHA'S DREAM

Among those who fell ill when *umbathalala* – the Spanish flu – reached the Ciskei region of South Africa in the autumn of 1918 was a Xhosa woman named Nontetha Nkwenkwe (*umbathalala* is Xhosa for 'disaster'). Regaining consciousness after her fever, she thought she had died and come back to life. People were bending over her, holding her hands and splashing water on her face. She began to recount a series of dreams that had come to her while she was ill.[13]

In one, she had seen an object wrapped in a dirty cloth hanging from the branch of a knobwood tree. A voice told her it was the Bible, but that it had rotted. When she asked Jesus for a piece of it so that she might testify to what she had seen, he refused. 'We have already given the Bible to people,' he said, 'but they have neglected it.' She was also told that people's sons were dying in the gold mines because they had stopped praying, and she was instructed to go to the places of the great chiefs and ask them if they were ready to be liberated and to work together to rule themselves. She was to preach to them, and to persuade them to look at their own lives and to stop blaming Europeans for their woes.

A voice then told her that *umbathalala* was a mere foretaste of the punishment that God would unleash on people for their sins. Judgement day had come. 'When I looked up at the heavens, I found them shaking just like the face of a cruel man. When the sun had risen above the earth in the east, it was red like burning charcoal. There was a person inside the sun and he was shaking his fists. And the heavens were coming together and I became afraid and cried. And there was a voice that said to me I should not cry but pray.' It had fallen to her to lead her people out of the ruins of their old society and towards a new one.

At the time that Nontetha fell ill she was in her forties, the widowed mother of ten children. She lived in a township called Khulile, in lands that had once belonged to her ancestors. The Xhosa had warred with Dutch and British settlers throughout the nineteenth century, and though they had scored some spectacular victories, they were now paying dearly for their ultimate defeat. The Land Act of 1913 had restricted black South Africans' share of the country's total land area to a risible 7.3 per cent. Squeezed into reserves in the Ciskei and Transkei (which are separated by the River Kei), many Xhosa found they were no longer able to live solely off the land. The men were forced to migrate to find work, leaving the women to run the home and family alone for six or nine months of the year. Nontetha's husband, Bungu Nkwenkwe, had worked first at the Kimberley diamond mines, and then at Saldanha Bay, an industrial area north of Cape Town, where he had died.

Although Nontetha was illiterate, she commanded respect in her community as an *ixhwele*, someone who knew the medicinal properties of plants. Part of the role of an *ixhwele* was to interpret events, especially traumatic ones, and there had been plenty of those in the past century. The Ciskei had known wars, famines, floods, even plagues of locusts – many of them in living memory. It was in the grip of a severe drought in the latter half of 1918, and then *umbathalala* arrived, coming down the train tracks with terrorised men fleeing the mines. In Nontetha's area roughly one in ten died of the flu – more than 10,000 people – and almost every family was touched (she herself lost a child to it).

Witnesses described bodies lying where they fell, in the bush or at the roadside. An eerie hush descended on the country. A missionary reported 'cattle, sheep and goats straying, unherded, and no one to secure the milk, so badly needed, from the uneasy cows'. With so many sick, crops went unplanted or unharvested, exacerbating the hunger. In the circumstances, when Nontetha recounted her dreams, people listened. Some of those who heard her laughed, but others took her seriously. 'It should be realised that Xhosa people attach great importance to dreams,' wrote the Xhosa poet James Jolobe in his 1959 poem 'Ingqawule'. 'They are the means of mediation between this world and the next.' Those who listened came back to hear her again, and gradually, she acquired a following. Nontetha had become a prophet.

She preached outdoors, wearing a white robe and headdress and armed with an *umnqayi*, a black ceremonial stick carried by senior married women. To the Xhosa, white signified healing and transformation; to Christians, purity. Nontetha's message appealed to both too, combining as it did both biblical and Xhosa references. The knobwood tree, for example, is known among Xhosa for containing a substance that, when rubbed on a breastfeeding woman's nipples, induces her baby to suckle. In her dream Nontetha had seen the Bible hanging from it, the implication being that people who had turned away from God must be induced to return to him. She herself had always mixed traditional and western clothes, and though she belonged to no church, all her children had been baptised Methodists, and she had great respect for missionary education.

Nontetha wasn't the only prophet to emerge at that time, and though she had no links with political organisations, many of the others did. They were responding to deep insecurities in the populations they spoke to, and to a general yearning for a better world. The years 1917 to 1920 saw a series of strikes on the Rand, as miners were drawn into the trades union movement and a nascent organisation called the African National Congress (ANC). A Zulu woman named Josephina started prophesying during the pandemic, and by 1923, she was sharing a platform with the ANC on the Rand, and predicting plagues of locusts with human heads and scorpion tails.

Afrikaners had their own insecurities. Accounting for more than half the country's white population, they resented the domination by an English-speaking minority of industry, the army, the arts and most other areas of South African life. Memories still rankled of the Anglo-Boer war at the turn of the century, in which 26,000 Afrikaners had died, and of another failed rebellion against the British in 1914. In 1916, an Afrikaner woman named Johanna Brandt had predicted a great plague that would usher in a new and better society. Two years later, her prophecy came true. But Afrikaner losses to the flu, though tiny compared to those sustained by the black population, only sharpened their sense that the volk was endangered.

The authorities became aware of Nontetha's activities in 1922. Many of her messages – warnings against the dangers of witchcraft and alcohol, for example – would have appealed to them, had they been inclined to listen. But by then they were extremely wary of new religious movements, or political movements masquerading as religious ones, as they tended to see them. A few years earlier at Bulhoek, less than 200 kilometres from Khulile, thousands of followers of a Christian movement called the Israelites had gathered to await the end of the world. When it didn't happen as their prophet had predicted, they stayed on anyway. More or less diplomatic attempts to disperse them failed, the police resorted to violence and more than 160 Israelites died in the ensuing massacre. Those in authority viewed Nontetha through the lens of Bulhoek. They saw her as subversive and anti-white – grounds enough to arrest her. Declaring her unfit to stand trial by reason of insanity, they committed her to a psychiatric hospital in Fort Beaufort, eighty kilometres from Khulile.

She was diagnosed with dementia praecox and released soon after her hospitalisation on condition that she refrain from preaching. Local magistrates called on Xhosa elders to enforce the prohibition, but they were unable to – partly because her female disciples defied them. Nontetha preached and her followers came to hear her. She was arrested again and readmitted to Fort Beaufort, but this didn't discourage her followers who, much to the irritation of the hospital administration, kept up a near constant presence there. So in 1924, Nontetha was moved to the notorious Weskoppies lunatic asylum in Pretoria, almost 1,000 kilometres from her home. There she witnessed at first hand the dark underbelly of the migrant labour system, because Weskoppies was a sort of holding camp for those who had gone to make their fortunes in the mines, and whose minds had been broken by them.

Nontetha found herself in an impossible situation. Each time she insisted that she had been inspired by God, her doctors took it as confirmation of their diagnosis and a reason not to discharge her. Her followers did not forget her, however, nor would they accept that she was mad. In 1927, a group of them walked for two months to reach Pretoria, and they were allowed to see her – though their requests for her release were rejected. Later 'pilgrimages of grace' were turned back, however, and in 1935 Nontetha died of cancer, cut off from her community and probably in pain, never having left the hospital. She was buried in an unmarked grave, the authorities having refused to hand her remains over to her devotees.

In 1948, the right-wing National Party came to power and imposed apartheid on the country (it also sought to promote Afrikaner culture and to improve Afrikaner health). The ANC was banned in 1960, and the ban remained in place until 1990. In post-apartheid South

Africa, an American historian named Robert Edgar was able to pursue enquiries that had previously been blocked, to locate Nontetha Nkwenkwe's remains. He tracked her to a pauper's grave in Pretoria, which she shared with an unnamed man. The man had been buried in a coffin – albeit a rough box – while she had no such covering, so as his coffin had disintegrated, their bones had mingled. At her exhumation, therefore, the remains of two probable strangers had to be disentangled, before hers could be returned to Khulile and reburied in the presence of her family and followers. Several thousand people attended her funeral on 25 October 1998, on the eightieth anniversary of 'Black October'.

17

Alternate histories

'Painful readjustment, demoralization, lawlessness: such are the familiar symptoms of a society recovering from the shock of the plague.'[1] When the historian Philip Ziegler wrote those words, he was describing the fallout from the Black Death, but they also apply to the Spanish flu. One in three people on earth had fallen ill. One in ten of those – perhaps as many as one in five – had died. If humanity had shown resilience, that was only evident from afar – at the population level. As soon as you came closer, as soon as you could make out individuals, it was impossible not to be struck by the price people paid for that recovery.

Families were forced to recompose themselves. From a distance of a hundred years, everything seems to have happened as it should have, since many of us are alive today because of that enforced game of musical chairs. We trace ourselves in a straight line back to those of our ancestors who survived. But they, looking forward, might have imagined other futures, other families. Renovating his home in 1982, Anders Hallberg, a farmer living near Sundsvall, Sweden, found a packet of letters bricked up inside a wall. The house had been inhabited by his family for generations. When he opened the packet, he realised they were love letters exchanged by his grandfather, Nils, and Nils's first wife, Clara. She was known as 'the beautiful Clara' in the village, and Nils had loved to play the piano for her. In one later, dated 17 January 1918, Clara wrote: 'My own beloved Nils . . . I'm longing to hug you and tell you how much I've missed you. My train arrives at five o'clock on Saturday. I'm sending you a thousand warm greetings and kisses. Your Clara. PS I spoke to Engla today, she sends her regards.'[2] Nils and Clara were married in August 1918, but the following April Clara died of the Spanish flu. Nils remarried a few years later – Engla – and in 1924 Engla gave birth to a son, Anders' father. But Nils never touched his piano again, and clearly, he wasn't able to destroy the letters.

'Fela would have been the prettiest,' wrote Jarosław Iwaszkiewicz in *The Maids of Wilko*, a story that the Polish director Andrzej Wajda made into a film in 1979, in which Fela's death from Spanish flu haunts her five surviving sisters. For decades after, people had a chronic sense of what might have been – of 'alternate histories'. So many had died, and so often death had appeared to strike at random. What if it had struck differently? It was a preoccupation of the survivors, perhaps even a kind of survivor's guilt. The elderly parents who had lost grown-up children bore it in silence, as was expected of them, and so Schiele's painting *The Family* is celebrated, while we know nothing about the grief of his mother, Marie, who outlived him by seventeen years.

In this global reshuffle, some fell through the cracks: long-term invalids, including melancholics, who could no longer work, and who were as misunderstood as (though probably more numerous than) war veterans suffering from the 'Flanders blues'; widows who had no hope of finding another husband; orphans nobody wanted. Because the flu had targeted those aged twenty to forty, many dependants found themselves deprived of their breadwinners. Some were caught in a very fragile, very threadbare safety net. Among them were the lucky beneficiaries of life insurance policies: the US life insurance industry paid out nearly $100 million in claims after the pandemic – the equivalent of $20 billion today. Others had been named in wills. Upon the death from flu of one German immigrant to America, for example, his widow and son received a sum of money. They invested it in property, and today the immigrant's grandson is a property magnate purportedly worth billions. His name is Donald Trump. Most had a less rosy future to look forward to, however. One Swedish study found that for each flu death, four people moved into the poorhouse.[3]

A person who was accepted into a public poorhouse in Sweden at that time received food, clothing, medical care and their funeral costs, but was declared legally incompetent.

Such studies are rare. Most of the information that survives about these casualties is anecdotal – and even then, their voices are faint. The plight of the orphans is especially troubling. Though there are no firm data on them, and although fewer children were born during the war than in peacetime, the fact that the flu targeted those in the prime of life – including young parents – suggests that there may have been a very large number of them. Adoption was not organised as it is now, and many would have been absorbed by the extended family or made wards of state. Ante Franicevic was born in a small village on the Neretva River in Croatia, one of four young siblings who lost both their parents and their paternal grandmother to flu within a matter of days. They were brought up by a series of uncaring relatives, until Ante, coming of age, decided with a friend to leave Croatia and make a new life for himself in Africa. They arrived in Northern Rhodesia (Zambia) soon after the Anglo American mining company had moved in to develop mines along the Copperbelt. The area where they found themselves was practically uninhabited, and to begin with they camped in snake-infested jungle, but their fortunes rose with the company's – particularly as demand for copper grew in the run-up to the Second World War. Ante worked for Anglo American for twenty-five years, married, raised a family and retired, comfortably off, to South Africa.

When there was no one to take them in, the orphans' prospects were bleak indeed. In the 1970s, an elderly German woman, Pauline Hammer, wrote to tell Richard Collier that she had lost both her parents to flu in 1919. Her eighteen-year-old sister had tried to keep the family together – eight-year-old Pauline, two other siblings and a foster brother – 'but after nine months or so we had to break it up'. She didn't explain what happened to them, only that losing her parents had cast a shadow over her life. Were the consciences of some governments pricked? It is possible, though difficult to prove, that the existence of flu orphans contributed to France's legalisation of the adoption of minors in 1923, and to Britain's legalisation of adoption three years later, after a century of fruitless campaigning. Those laws benefited millions of children, but they came too late to help many of the orphans of the Spanish flu.

AIDS has created millions of orphans, Ebola thousands. Welfare organisations report that these orphans are more likely to drop out of school, to be malnourished, to live on the streets, to be exploited by adults, and to be drawn into prostitution and crime. That is the situation today, and in 1918 it was certainly no better. An estimated 500,000 children were orphaned in South Africa alone, during Black October. The South African government, along with the police, the post office, the railways and certain religious institutions, launched an ambitious programme of orphanage building, but it catered mainly to the white minority. Very little was done for the hundreds of thousands of black or coloured orphans who, if they weren't taken in, often ended up as indentured labourers – domestic servants or farmhands – or vagrants.

In 1919, indicting one coloured 'flu remnant' for theft, a Cape Town prosecutor painted a vivid portrait of the accused: 'He has no home, and does not know what has become of his parents. He does not know his age or his proper name, and has no surname, so far as he knows. He and others sleep under the pier, in the old boxes, and in railway compartments, first-class preferred, when the opportunity offers. He looks half starved and eats garbage, or whatever he can get hold of, and says he has never been to school.' He was 'one of dozens of boys his age who roam the city and sleep anywhere'. The presiding magistrate found the boy guilty and sent him to a reformatory for four years.

Thus troublesome elements were brushed aside, in societies' march to recovery. New babies were born – a record number of them, in the 1920s – and populations replenished themselves. Some countries, at least, saw an economic rebound too. In America, industrial output and business activity took a serious hit in 1918 (with the exception of businesses specialising in healthcare products), due to the flu, but when economists Elizabeth Brainerd and

Mark Siegler looked at state-by-state flu mortality rates and compared them to estimates of personal income for the following decade, they found a striking correlation: the higher the death rate, the higher the growth in per capita income throughout the 1920s. This was not new wealth, but it was an indication of the capacity of a society to bounce back after a violent shock.[4]

Not all communities recovered. The island nation of Vanuatu is today home to over 130 local languages that are spoken in addition to English, French and the national language Bislama, making it the most linguistically dense country in the world (each local language has between 1,000 and 2,000 speakers, on average). Parts of the Vanuatu archipelago experienced 90 per cent mortality during the Spanish flu, and that epidemic – along with others of smallpox and leprosy that swept over the islands in the early 1900s – pushed around twenty local languages to extinction. The population is still recovering from that catastrophic collapse, but those twenty languages – and the cultures associated with them – are gone forever.[5]

Some have blamed the social ills that afflict many small-scale societies today on epidemics including the Spanish flu (though contact with outsiders changed their lives in many more ways besides the introduction of new diseases). When Johan Hultin returned to Brevig Mission in 1997, to reopen the mass grave where the village's flu victims had been buried, he found it a sad, hopeless place – quite unlike the one he had visited in 1951. Back then, local people had still practised whaling and hunting and were self-sufficient; now they were dependent on welfare handouts.[6] Of course, whaling and hunting are hazardous pursuits, and Hultin's impression may have been wrong – the canny villagers may have chosen to take whatever money the government was offering in order to devote their time and energies to less dangerous but nevertheless fulfilling activities. The findings of a report by the Alaska Natives Commission suggest otherwise, however. Published three years earlier, it described Alaskans as a 'culturally and spiritually crippled people' who had become dependent on others to feed, educate and guide them.[7]

The commission placed some of the blame on epidemics that had caused the deaths of shamans and elders – the repositories of knowledge and tradition in native Alaskan cultures – while at the same time creating many orphans. In the early twentieth century, it was common practice to remove such orphans from their communities and place them in centralised institutions. The idea was that this would encourage them to assimilate into a larger, more diverse community and broaden their horizons. Instead, the report claimed, they experienced 'long-term cultural loss'. These problems, aggravated by competition with outsiders for natural resources and for work in local industries, had culminated in a situation where 'the social and psychological condition of Native people has varied inversely with the growth of government programmes intended to help them'. The more money the government pumped in, in other words, the more rates of alcoholism, crime and suicide soared in Alaska.

One of those who contributed to the 1994 report was Yupik elder Harold Napoleon. Two years after it appeared, while serving time at the Fairbanks Correctional Center for the drunken killing of his infant son, he wrote an essay entitled *Yuuyaraq*. Literally 'the way of being a human being', Yuuyaraq is the name of the world the Yupik traditionally inhabited – a world busy with animal and human spirits. Napoleon's essay was a lament for that lost world and an attempt to understand what had happened to his people. His thesis, based on his own experiences and those of his fellow inmates, was that the epidemics that had battered them for nearly two centuries had destroyed their culture and left them traumatised – so traumatised that they could not even talk about it. 'To this day *nallunguaq* remains a way of dealing with problems or unpleasant occurrences in Yupik life,' he wrote. 'Young people are advised by elders to *nallunguarluku*, "to pretend it didn't happen". They had a lot to pretend not to know. After all, it was not only that their loved ones had died, they also had seen their world collapse.'[8]

18

Anti-science, science

In 1901, Gustav Klimt shocked Viennese society when he unveiled *Medicine*, one of a series of works he had been commissioned to paint to decorate the ceiling of the Great Hall of the University of Vienna. The theme of the series was the triumph of light over darkness, but Klimt's painting placed a skeletal Death in the middle of a cascade of naked bodies – the river of life. His meaning was clear: when it came to the healing arts, darkness continued to triumph over light. The Ministry of Education refused to fix *Medicine* to the ceiling, and Klimt resigned the commission, saying that he wanted to keep the work for himself. Fearing that he intended to exhibit it abroad, the ministry claimed it was state property and sent agents to seize it. Klimt threatened them with a shotgun and they came away empty-handed.[1]

The artist had witnessed the deaths of his father, brother and sister, and lost his mother and another sister to insanity. Like many famous men and women – and many more who weren't famous – disease had blighted his life. Nor was he the only one warning the medical profession against hubris in the early twentieth century. In 1906, George Bernard Shaw wrote *The Doctor's Dilemma*, in which the eminent doctor Sir Colenso Ridgeon toys godlike with his patients' destinies (Sir Almroth Wright, on whom the character was based, is said to have walked out of a performance). But in Europe, the cradle of germ theory, they were swimming against the tide. Only after the Spanish flu did the backlash become generalised. On 28 October 1918, *The Times* of London muttered darkly about neglect and lack of foresight, and looked to make 'somebody answerable for the nation's health'. 'Science has failed to guard us,' stated the *New York Times*, a paper that spoke to one of the most enthusiastically scientific nations on earth.[2] 'No more dope!' clamoured the enemies of western medicine everywhere.

The hubris of the medical community was punished, at least in the industrialised world. The irregulars had all claimed higher cure rates than the regulars, and their followings now grew. Over the next two decades, as scientists argued over what had caused the Spanish flu, they flourished and acquired respectability – including the more respectable label, 'alternative medicine'. In the 1920s, in some US cities, a third of those who went to conventional doctors also went to alternative practitioners. Chiropractic reached mainland Europe at the beginning of that decade, and by the end of it the only continent where it wasn't practised was Antarctica. As for homeopathy, the man who had presided over the health of New Yorkers during the pandemic, and who as a surgeon and homeopath had a foot in both camps – Royal S. Copeland – legitimised it when, as a senator for New York, he made sure its pharmacopoeia was approved under the Federal Food, Drug, and Cosmetic Act of 1938.

The irregulars had a fundamentally different conception of health from Ilya Mechnikov. For Pasteur's lieutenant, nature was essentially disharmonious and in need of a helping hand – notably vaccination – to coax it into health. For them, disease was the result of a disruption of the natural harmony, and so was vaccination, to which they were fiercely opposed. Benedict Lust, the father of naturopathy, called germ theory 'the most gigantic hoax of modern times'.[3] As the irregulars gained in stature, some of their ideas penetrated the popular consciousness and were eventually embraced even by the conventionals. The most important of these was an emphasis on prevention that went beyond hygiene, to sport, body consciousness and diet. These ideas reached the masses with the blessing and encouragement of the elites, who saw in them a convenient way of distracting the lower classes from the dangerous allure of communism. Thus the King of Spain – the same Alfonso XIII whose high-profile case of flu had contributed to the pandemic's naming – gave his regal imprimatur to the Madrid Football Club in 1920, creating Real (Royal) Madrid FC, and turned football into a national pastime.

Back-to-nature movements had taken off in the nineteenth century, as an antedote to industrialisation, but they had been fairly elitist affairs. In the 1920s, movements such as *Lebensreform* (life reform) in Germany – which advocated vegetarianism, nudity and homeopathy – widened their remit, drawing in just those sectors of the population that had suffered worst from Spanish flu. In 1918, the Italian-Americans of New York, like the Jews of Odessa, had kept their windows firmly shut, believing that spirits or bad air caused disease. Now sunlight and fresh air became bywords for health, and by 1930 the concepts of nature and cleanliness had become firmly linked in people's minds. Counter-intuitively, after the war, a once vigorous anti-cigarette movement collapsed. Smoking had been encouraged among the troops as a least-worst substitute for other sins, but it had also been promoted as a prophylactic against Spanish flu. Now associated with positive attributes, it became fashionable. Women took it up.

Among the bitterest enemies of conventional medicine were Christian Scientists, who rejected almost all medical intervention. The pandemic over, they claimed that prayer alone had proved superior to conventional methods, and their following also grew rapidly at this time – both in their native US and abroad. New faith-healing movements were born. Philadelphia had had a particularly severe dose of the flu, and in October 1918 – just as the *New York Times* was trumpeting the failure of science – the mouthpiece of the Philadelphia-based Faith Tabernacle, *Sword of the Spirit*, published the testimonies of the 'healed' of the Spanish flu under the headline 'God's witnesses to divine healing'. That year, Faith Tabernacle established itself in Gold Coast (Ghana) – which had also had a bad case, losing an estimated 100,000 people in six months – and quickly spread to Togo and Ivory Coast. Faith Tabernacle was in decline in West Africa by the late 1920s, but it lives on in the African

Pentecostal movement, with its emphasis on divine healing and speaking in tongues.

Many Africans underwent a crisis of comprehension in 1918, since neither Christian missionaries – whom they associated with western medicine – nor their traditional healers could account for the scourge.[4] A new generation of prophets emerged who offered a different world view. Flu survivor Nontetha Nkwenkwe was one of them, in South Africa, and her story ended tragically in a clash with western medicine. But Africans weren't the only ones undergoing intellectual crises. 'Victorian science would have left the world hard and clean and bare, like a landscape in the moon,' wrote Sir Arthur Conan Doyle in 1921, 'but this science is in truth but a little light in the darkness, and outside that limited circle of definite knowledge we see the loom and shadow of gigantic and fantastic possibilities around us, throwing themselves continually across our consciousness in such ways that it is difficult to ignore them.'[5]

Conan Doyle, the British creator of that most scientific of detectives, Sherlock Holmes, stopped writing fiction after he lost his son to the Spanish flu, and devoted himself instead to spiritualism – the belief that the living can communicate with the dead. Spiritualism had been popular in the nineteenth century, but it enjoyed a resurgence after 1918, encouraged in part by Albert Einstein's description of time as a fourth dimension (if there were four, why couldn't there be more, some of which harboured restless spirits?). In 1926, Conan Doyle was invited to speak to the members of a scientific society at Cambridge University, who listened politely if sceptically to his description of ectoplasm as the material basis of all psychic phenomena.[6]

In general, the 1920s were a time of intellectual openness, of testing and trampling of boundaries. With the publication of his general theory of relativity in 1915, Einstein had introduced the notion of the subjectivity of the observer. Niels Bohr and Werner Heisenberg were arguing, within a decade of the Spanish flu, that there is no knowledge without uncertainty. Any scientist who had lived through the pandemic, especially if he remembered Émile Roux's insightful musings on those *êtres de raison* (organisms whose existence can be deduced only from their effects), realised that good science demanded open-mindedness, experimental rigour and a healthy dose of humility.

If such ideas were circulating at that time, some credit must go to the Pope. In 1919, nothing remained of the international scientific community that had thrived before the war. If an international science conference had been organised in that year, Germans and Austrians would have been excluded. The Vatican had annoyed both sides by declaring itself neutral in 1914, and in 1921, wanting both to re-establish peace and to re-ingratiate himself, Pope Benedict XV revived the moribund Lincean Academy – a forerunner of the Pontifical Academy of Sciences. He gave it the mission of restoring international scientific relations, seeing the quest for a disinterested truth as the perfect vehicle for dialogue, but he was choosy about his quests. Only 'pure' or experimental science qualified – physics, chemistry, physiology. The applied sciences that aimed to solve human problems were, in his view, subjective and thus liable to reproduce the tensions that had led to war in the first place.[7]

The entente nevertheless extended to all sciences, eventually, and by the 1930s, medical science had redeemed itself, to some extent. Virology was established as a discipline, the first flu vaccines were coming online, and Fleming had discovered penicillin while trying, and failing, to grow Pfeiffer's bacillus in a dish. By then, however, thanks to his success in publishing nature cure journals, American naturopath and wrestler Jesse Mercer Gehman had accumulated a larger fortune than the press baron William Randolph Hearst (whose mother, Phoebe, had died of Spanish flu). And the Nazis, in power in Germany, had appropriated the notion of nature as clean to legitimate purification of the German population, a project that culminated in the Second World War. As the Schutzstaffel (SS) retreated at the end of that war, they set fire to the Austrian castle where works from Vienna's Belvedere

Museum had been stored for safekeeping, Klimt's *Medicine* among them. All that remains of the work today are a few sketches and some poor-quality photos. The artist never discovered its fate, because he died in February 1918. He had suffered a stroke and caught pneumonia while in hospital. Some have suggested that his was an early case of Spanish flu.

19

Healthcare for all

If health authorities had learnt anything from the pandemic, it was that it was no longer reasonable to blame an individual for catching an infectious disease, nor to treat him or her in isolation. The 1920s saw many governments embracing the concept of socialised medicine – healthcare for all, free at the point of delivery.

You don't put a universal healthcare system in place in one fell swoop. Such a system takes time to develop, and to become truly universal. The first and most important step is to work out how you are going to pay for it. Germany was a pioneer in this, Chancellor Otto von Bismarck having set up a national medical insurance programme in 1883. Under this state-funded, centrally administered scheme – which lives on in spirit in the country's modern healthcare system – Germans could expect to receive treatment and sick pay for up to thirteen weeks. Britain and Russia established insurance systems in the 1910s, but it wasn't until the following decade that most countries in western and central Europe followed suit.

Once you have your funding in place, the next step is to reorganise the way you provide your healthcare. In Germany at the time of the Spanish flu, healthcare was fragmented. There was no national health policy, though the idea had been floated in 1914, and doctors worked either on their own account, or were funded by charities or religious institutions – a pattern repeated all over the industrialised world. In 1920, a social hygienist in Baden named Ernst Künz suggested a root-and-branch reform, whereby the government would train and fund district physicians, and health councils would be elected at every level of the country's administration.[1] Künz's proposal was ignored – perhaps, some have suggested, because if they had acknowledged the need for change, German physicians would have been admitting to their failure to manage the Spanish flu, and they weren't yet ready to do that.

In 1920, therefore, Russia was the first to implement a centralised, fully public healthcare system. It wasn't universal, because it didn't cover rural populations (they would finally be included in 1969), but it was a huge achievement nevertheless, and the driving force behind it was Vladimir Lenin. He was well aware that, although the revolution had succeeded, it had done so at the cost of the near annihilation – due to famine, epidemics and civil war – of the working class. Doctors feared persecution under the new regime (the Bolsheviks weren't fond of intellectuals), but Lenin surprised them by soliciting their involvement at every level of the new health administration, and in the early days, this placed particular emphasis on the prevention of epidemics and famine.

The official Soviet vision of the physician of the future was spelled out in 1924, when the government called on medical schools to produce doctors who had, among other things, 'the ability to study the occupational and social conditions which give rise to illness and not only to cure the illness, but to suggest ways to prevent it'.[2] Lenin realised that medicine should be not only biological and experimental, but also sociological, and it was around the same time that epidemiology – the science of patterns, causes and effects in disease, that is the cornerstone of public health – received full recognition as a science.

Epidemiology requires data, and in the years following the pandemic the reporting of health data became more systematic. By 1925, all US states were participating in a national morbidity reporting system. The early warning apparatus that had been so lamentably lacking in 1918 started to take shape, and public health officials also began to show more interest in a population's 'baseline' health. The first national health survey was carried out in America in 1935 – eighteen years after the 'horrible example', when the mass examination of army draftees had revealed shocking levels of preventable or curable illness and deformity.

Governments beefed up their epidemic preparedness. Nowhere was this feat more remarkable than in China, where – in the years since the Manchurian plague outbreak of 1911 – Wu Lien-teh had, almost singlehandedly, put in place the foundations of a modern health system. In 1912 he had set up the North Manchurian Plague Prevention Service. The following year, dissection was legalised in medical schools, and in 1915, the National Medical Association was created to promote western medicine in China, with him as its first secretary. After Chiang Kai-shek seized power from the warlords, his regime centralised the collection of health data, and in 1930 a National Quarantine Service was set up in China. Under its first director, Wu, this organisation oversaw quarantine arrangements at all China's major ports, and sent regular epidemiological reports back to the League of Nations in Geneva. Within two years of crowning himself Shah of Persia, meanwhile, General Reza Khan had snatched back quarantine services on the Persian Gulf from the British – though not without a struggle – and between 1923 and 1936, his government increased fiscal allocations to the country's sanitary infrastructure twenty-five-fold.[3]

As more disease data became available, and as more people were drawn into the universal healthcare 'net', epidemiology's scope broadened. To begin with, it had been narrowly focused on infectious disease, but it soon encompassed non-communicable or chronic diseases too, and by 1970 epidemiologists were interested in any health-related outcome – even homicide. That evolution reflected both scientific progress and demographic change, as heart disease, cancer – and more recently the dementias – overtook infectious diseases as the biggest killers.

When Britain set up its National Health Service in 1948, pneumonia, TB, polio and venereal disease still killed large numbers of people, and one in twenty babies died before the age of one (tenfold more than today). Medical science wasn't what it is now, but it had nevertheless made great strides since 1918: there were modern antibiotics, and from 1955, a polio vaccine. This is why the NHS and similar systems were so transformative. Poor people who had previously received no medical care at all, who had relied on their own, sometimes dangerous folk medicine or on the charity of doctors, could now be cured of many of their illnesses. The elderly were among those who saw the biggest change, since many of them had been condemned to ending their lives in neglected 'back wards' or workhouses. The NHS pioneered the development of geriatric medicine in Britain.

Many of us take free healthcare for granted today, so it's easy to forget that the concept was extremely unpopular in some quarters, in the 1940s. Doctors tried to block the NHS for two years prior to its birth, considering it a threat to their income and their independence. It was seen as synonymous with socialism – a 'socialist plot' – and at one point Winston Churchill of the Conservative Party attacked the Labour health minister Aneurin Bevan in the House of Commons, calling him 'a curse to his country'. Indeed, fears of a 'socialist plot' are the reason Americans still don't have a universal healthcare system today. Instead, employer-based insurance systems began to proliferate in that country from the 1930s on.

Many countries created or reorganised health ministries in the 1920s. This was a direct result of the pandemic, during which public health leaders had been either left out of Cabinet meetings completely, or reduced to pleading for funds and powers from other departments. Now they had a seat at the high table, and thus, increasingly, public health became the responsibility of the state. At the same time, politicians realised that public health measures gave them a means of extending their influence over populations. Health became political, and nowhere more so than in Germany.

Though Ernst Künz's proposal of reform was ignored, the emphasis in German healthcare did gradually shift from private practice to public health under the Weimar Republic (1919–33), and by the time the Nazi Party came to power German doctors were used to cooperating with the government in the provision of medical care. Eugenics, of course, had long been a powerful current of thought, but in 1930s Germany, eugenic theory – as promulgated by the Nazis – became mainstream medical practice.

One of the first Nazi laws to be passed, in 1933, was the Law for the Prevention of Offspring with Hereditary Diseases – also known as the 'sterilisation law' – whose aim was to prohibit persons defined as genetically inferior from reproducing. 'Genetic health courts' made up of judges and doctors – with the doctors acting as 'advocates of the state' – made decisions about the forcible sterilisation of such individuals, in sessions from which the public was barred, and sometimes in less than ten minutes. A subsequent expansion of the law allowed them to order abortions up to the sixth month of pregnancy.[4]

The state of a nation's health came to be seen as an index of its modernity or civilisation. As disease surveillance improved, and health problems in the colonies of Africa and Asia became more visible, they became an embarrassment to colonial powers. At the same time, the indigenous inhabitants of those colonies became resentful of their own condition and blamed the colonisers for failing to provide adequate healthcare. They looked longingly to Russia and its system of universal coverage. The capitalist west had to come up with its own solution, and often that solution was furnished by the Rockefeller Foundation.

The Rockefeller Foundation was the philanthropic offshoot of Standard Oil, and it had been set up in New York State in May 1913 by that company's owner, John D. Rockefeller, his philanthropic advisor, Frederick Taylor Gates, and his son, John D. Rockefeller Jr. The foundation's International Health Division, created six weeks later, would become one of the most important players in international public health between the wars, helping to fight disease, not only in many colonies and newly independent states, but also in western Europe. In 1922, for example, it signed a deal with the Spanish government that put in place the building blocks of a modern health system in that country. It was also influential in helping Wu overhaul medical education in China – notably through the Peking Union Medical College, which it financed.

Rockefeller wasn't alone. The Pasteur Institute also spread its wings in those years, and in 1922 it set up an outpost in Tehran – the direct outcome of conversations between Émile Roux and the Persian delegates to the Paris peace conference, who had been traumatised by the devastation the Spanish flu had wrought in their country. In the immediate post-war period, when Europe was ravaged by epidemics – not only flu, but also typhus and TB – religious bodies organised humanitarian aid in affected areas, while the Save the Children Fund was set up in 1919 to provide relief for emaciated, disease-ridden Austrian and German children – victims of the war and the Allied blockade.

In the context of all this well-intentioned but relatively uncoordinated activity, a need was perceived for a new kind of international health organisation. The Paris-based International Office of Public Hygiene had been set up in 1907, with the blessing of twenty-three European states, but its function was mainly to collect and disseminate information regarding infectious disease, not to implement public health programmes. Something more proactive was now required, and in 1919, with the support of the Geneva-based International Committee of the Red Cross, an international bureau opened in Vienna with the express mission of fighting epidemics.

It was here, at the international level, that the two opposing forces shaping public health – socialisation and politicisation – now clashed. No sooner had the anti-epidemic bureau opened, than nations began squabbling over whether the defeated powers should be included in it, and anti-Semitic elements started lobbying for Jewish refugees to be quarantined in eastern European concentration camps. (The term 'concentration camp' wasn't new even then, having been used twenty years earlier to describe camps the British built to accommodate Boer women and children during the Second Anglo-Boer War; intended as humane shelters, they were soon overrun by disease.) Questions were also raised over the German POWs who were still in Russia: if there were Bolshevik agitators among them, should they be allowed to return home?

Eglantyne Jebb, the British founder of Save the Children, stood out in this debate for her insistence on inclusiveness – even of Bolsheviks. And it wasn't just the anti-epidemic bureau that was hijacked for political ends, or perceived to be. Rockefeller was suspected by some of practising neocolonialism under the guise of philanthropy. The foundation saw its mission as bringing American-style enlightenment to 'the depressed and neglected races', and it maintained close ties with businessmen and missionaries in the countries to which it delivered that enlightenment (its reputation would later be tarnished by its involvement in Nazi eugenics programmes).

In the early 1920s, the League of Nations established its own health organisation, and this – along with the anti-epidemic bureau, the older Pan American Health Organization and the Paris-based organisation – was the forerunner of today's World Health Organization (WHO). When both the League of Nations and its health branch collapsed in 1939, at the outbreak of the Second World War, it sent a clear message to the architects of the future WHO: the new organisation should not depend for its survival on that of its parent body, the United Nations. When the WHO was inaugurated in 1946, therefore, it was as an independent institution. By then, eugenics had fallen from grace, and its constitution enshrined a thoroughly egalitarian approach to health. It stated, and still states, that 'The enjoyment of the highest attainable standard of health is one of the fundamental rights of every human being without distinction of race, religion, political belief, economic or social condition.'

20

War and peace

Erich Ludendorff, the general who led the German war effort, thought that the Spanish flu had stolen his victory. There has been plenty of 'what if?' speculation regarding the First World War. What if Herbert Asquith's Liberal government had kept Britain out of it, as it so nearly did in 1914? What if America hadn't stepped in three years later? What if Fritz Haber hadn't discovered a way of manufacturing ammonia from nitrogen and hydrogen gases, allowing the Germans to continue making explosives despite the Allied naval blockade that prevented them from receiving shipments of saltpetre? Things happened the way they did because of a host of complex, interacting processes, and to try to pull one free of the mix risks misleading. Still, Ludendorff's claim merits closer examination, if only because he's not the only one to have made it. It has been repeated, in this century, by academics who study wars for a living.

When the Central Powers launched their spring offensive in late March 1918, they had the upper hand. The collapse of the Eastern Front had released large numbers of battle-hardened troops who were now retrained in such modern tactics as infiltration of enemy lines (these were the agile stormtroopers). Though food was scarce both at home and in the trenches, due to the blockade, the Germans felt they had reached a tipping point, and they were optimistic. Allied morale, on the other hand, was low. They were overstretched in terms of manpower and weary from years of failed offensives against the other side. The previous autumn, mutiny had broken out at Étaples, and been brutally put down.

The first phase of the offensive was successful, and by early April the Germans had pushed the Allies back more than sixty kilometres. On 9 April, they launched a second phase, Operation Georgette, and made further gains. In grim mood, the British commander-in-chief Sir Douglas Haig urged his men to 'fight it out' to the last. But then Georgette began to run out of steam, and it was called off at the end of April. On 27 May, a third phase began – Operation Blücher – but Blücher was faltering by early June. A successful French counter-attack halted the Kaiser's Battle in July, and starting in August the Allies launched a series of offensives that pushed the Central Powers out of France and brought the war to an end.

By June, the Central Powers had outrun their supply lines, and they too were exhausted. But as this timeline shows, things started going wrong for them earlier – around the middle of April. It was then that the flu first appeared in the trenches. Both sides sustained heavy casualties from the disease, but Ernst Jünger, a German stormtrooper who had been sent with his company to defend a small wood twenty kilometres south of Arras (the British called it Rossignol Wood, the Germans Copse 125), felt that his side was hit harder. Several of his men reported sick each day, he recalled later, while a battalion that had been due to relieve them was almost 'wiped out'. 'But we learned that the sickness was also spreading among the enemy; even though we, with our poor rations, were more prone to it. Young men in particular sometimes died overnight. And all the time we were to be battle-ready, as there was a continuous cloud of black smoke hanging over Copse 125 at all times, as over a witches' cauldron.'[1]

Most historians are reluctant to suggest that the flu determined the victor of the war, though they do agree that it accelerated the end of hostilities. Two have broken ranks, however, and suggested that the flu 'punished' the Central Powers more severely than the Allies, thereby biasing the outcome. Military historian David Zabecki agrees with Jünger's claim that malnutrition in the German ranks exacerbated the flu among them,[2] while political scientist Andrew Price-Smith argues that the lethal autumn wave may have been the last straw for the tottering Austro-Hungarian Empire.[3] Ludendorff may have seen the writing on the wall, where Germany was concerned: towards the end of September, he suffered something resembling a nervous breakdown, and his staff called a psychiatrist.

Conditions were very bad in Central Europe by the autumn of 1918, though the true gravity of the situation didn't become apparent to those beyond its borders until after the war had ended. Writer Stefan Zweig had a foretaste of it when, travelling back to his native Austria in the months following the armistice, his train was stopped at the Swiss border. There he was asked to leave the 'spruce, clean' Swiss carriages and step into the Austrian ones:

> One had but to enter them to become aware beforehand of what had happened to the country. The guards who showed us our seats were haggard, starved and tatterdemalion; they crawled about with torn and shabby uniforms hanging loosely over their stooped shoulders. The leather straps for opening and closing windows had been cut off, for every piece of that material was precious. Predatory knives or bayonets had had their will of the seats, whole sections of the covering having been rudely removed by such as needed to have their shoes repaired and obtained their leather wherever it was to be had. Likewise the ashtrays were missing, stolen for the sake of their mite of nickel or copper.[4]

The British economist John Maynard Keynes issued a warning about the dire situation in the defeated countries in his book *The Economic Consequences of the Peace* (1919). 'For months past the reports of the health conditions in the Central Empires have been of such a character that the imagination is dulled, and one seems almost guilty of sentimentality in quoting them,' he wrote, before going on to quote a Viennese newspaper: 'In the last years of the war, in Austria alone at least 35,000 people died of tuberculosis, in Vienna alone 12,000. Today we have to reckon with a number of at least 350,000 to 400,000 people who require treatment for tuberculosis . . . As the result of malnutrition a bloodless generation is growing up with undeveloped muscles, undeveloped joints, and undeveloped brain.' We know that the tubercular were particularly vulnerable to the flu, and if it is true that Switzerland and France received the second wave of the pandemic from the east, as some sources suggest, then Austro-Hungary may have been exposed to it for longer than those countries, and sustained proportionally greater losses. It seems at least possible, therefore, that there is some merit to Ludendorff's claim, and that the flu favoured the Allies.

What about the peace – did flu have a hand in that too? Some historians think so. The third wave struck Paris in the middle of the peace process, and delegates involved at every level of the difficult and protracted negotiations were affected by it, directly or indirectly. Wellington Koo of the Chinese delegation, who was fighting for the return of Shantung to China – and ultimately, to restore China's dignity – lost his wife to it. T. E. Lawrence, Lawrence of Arabia – who was accompanying the Arab delegation led by Prince Faisal (later King Faisal I of Iraq) – absented himself briefly to go to England on hearing that his father was dying of the flu. Arriving two hours after Lawrence Snr's death, he turned around and came straight back, not wanting to be away too long from discussions about the future of the predominantly Arab lands that had until recently belonged to the Ottoman Empire.[5] David Lloyd George had recovered from his bout of the previous autumn, but the French prime minister, Georges Clemenceau, was racked with 'colds' throughout March and April. Clemenceau had survived an assassination attempt in February, and though these might have been sequelae of the bullet lodged behind his shoulder blade, they might also have been the Spanish flu.

Perhaps the most significant victim of the flu in Paris that spring, however, was the American president, Woodrow Wilson.

He soldiered on, but observers noticed that this usually calm and deliberate man became, on occasion, forgetful, irascible and quick to judge (unfortunately his closest advisor, Edward House, also had a bad case). Wilson had an underlying neurological weakness and may have been suffering transient ischaemic attacks – mini strokes – for years.[6] Some contemporary neurologists who have studied his case claim that the flu triggered further mini strokes that spring (others disagree, and retrospective diagnosis is a notoriously tricky affair). If so, did they affect the outcome of the negotiations?

Wilson was certainly a key player in those negotiations. Armed with his Fourteen Points, he fought an often lonely battle for a moderate peace and a League of Nations against his more vengeful European counterparts. But one of his recent biographers, John Milton Cooper Jr, doesn't think his parlous state of health that spring *did* have a lasting impact. With the important exception of Shantung – which was conceded to Japan in exchange for a pledge to join the League, much to Chinese rage and Wilson's chagrin – Cooper says that he essentially achieved all his aims in Paris. When it came to the reparations to be paid by the defeated nations, arguably the most damaging product of the peace process because of the humiliation and hardship it inflicted on Germany, the delegates agreed only on the principle, not on the actual amounts to be paid. Those were arranged later, by representatives of all the nations that had ratified the treaty – and as it turns out, the US was not among them.[7]

But if the experts can't agree on Wilson's neurological state in the spring of 1919, they do reach a degree of consensus when it comes to the massive stroke he suffered the following October. His earlier bout of flu certainly did contribute to that, they believe. That stroke left an indelible mark both on Wilson (leaving him paralysed down the left side) and on global politics, in Cooper's view, because it rendered him unable to persuade the US government to ratify the Treaty of Versailles, or to join the League. Germany was forced to pay punitive reparations, stoking its people's resentment – something that might not have happened had the US had a say in it. By turning Wilson into the greatest obstacle to his own goals, the Spanish flu may therefore have contributed, indirectly, to the Second World War.

Away from the peace process, the flu shaped other important political events. In March 1919, Yakov Sverdlov, chairman of the All-Russian Central Executive Committee, caught the disease and died within a week. A small, imperious man with a serious voice, who liked to dress from head to toe in black leather, he had been Lenin's right-hand man since the latter was shot and badly injured in an assassination attempt the previous August. Leon Trotsky reported that Lenin phoned him at the war commissariat with the news of Sverdlov's death: '"He's gone. He's gone. He's gone." For a while each of us held the receiver in our hands and each could feel the silence at the other end. Then we hung up. There was nothing more to say.'[8] Sverdlov was buried on Red Square in the Bolsheviks' first major state funeral. Sverdlov replacements came and went – all of them lacking his formidable energy, all of them unequal to the enormous task of building a communist state from scratch – until in 1922, Joseph Stalin stepped into the role.

Two months before he fell ill in May 1918, the Spanish king, Alfonso XIII, narrowly avoided a coup. Having risen from his sickbed, he managed to cobble together a new coalition government by pleading with opposing factions to come to the table, and threatening to abdicate if they didn't. It was his last-ditch attempt to save the *turno pacífico*, the compromise that had brought to an end the turmoil of the nineteenth century by ensuring that liberals and conservatives took turns to rule in governments chosen by him. Some have argued that had the king not recovered, or had his convalescence taken much longer, Spain would have become a dictatorship a few years earlier than it did. As it was, a coup led by General Miguel Primo de Rivera in 1923 ushered in a period of dictatorship for which Spaniards had arguably shown their appetite in their calls for a sanitary dictatorship in 1918. They were desperate for a strong hand at the helm, someone to steer them out of their backwater and back into the main European current.[9]

The autumn of 1918 saw a wave of workers' strikes and anti-imperialist protests across the world. The disgruntlement had been smouldering since before the Russian revolutions of 1917, but the flu fanned the flames by exacerbating what was already a dire supply situation, and by highlighting inequality. It hurled a lightning bolt across the globe, illuminating the injustice of colonialism and sometimes of capitalism too. The eugenically minded who had noticed how badly the underclasses had suffered tended to blame their inferior stock. But the underclasses had also noticed the disparities, and they interpreted them as proof of their own exploitation at the hands of the better-off. In the French colony of Senegal, for example, it did not go unremarked that colonial doctors had prescribed champagne for Europeans, and wine for natives.[10] Revolution broke out in Germany in November 1918, in the midst of the autumn wave, and even well-ordered Switzerland narrowly avoided civil war, after left-wing groups blamed the high number of flu deaths in the army on the government and military command.

Western Samoa experienced one of the highest flu-related mortality rates in the world, losing more than one in five of its population after the infected New Zealand steamer the *Talune* arrived at its capital Apia in November 1918. The catastrophe exacerbated indigenous resentment against the islands' New Zealand administration, and the 1920s saw a revival of the Mau – a non-violent protest movement that had been mobilised against the islands' previous occupiers, the Germans (New Zealand seized the islands from Germany at the outbreak of the war). In 1929, during a peaceful demonstration in Apia, police tried to arrest the Mau's leader, High Chief Tupua Tamasese Lealofi. A struggle ensued and they fired into the crowd, killing Tamasese and ten others. The Mau's popularity only grew after that, and following a number of failed attempts, Western Samoa (now Samoa) finally obtained its independence in 1962. Neighbouring American Samoa remains a US territory.

In Korea, as we've said, ethnic Koreans were twice as likely to die from the Spanish flu as their colonial masters, the Japanese. The flu-related death rate in Egypt, meanwhile, was roughly twice that in Britain. In March 1919, Koreans rose up in an independence movement that the Japanese quickly crushed (Koreans finally gained their independence after the Second World War), and in the same month Egyptians and Sudanese people revolted against their British 'protectors' – a revolution that would lead to Egypt gaining its independence in 1922. By March 1919, meanwhile, tensions in India had reached breaking point, in large part due to the flu. In that country, however, they wouldn't come to a head until the following month.

GANDHI AND THE GRASS ROOTS

Throughout the summer of 1918, Mahatma Gandhi was busy recruiting Indian troops to the British war effort. By the autumn he was worn out, and while at his ashram on the outskirts of Ahmedabad, he suffered what he thought was a mild attack of dysentery. He made up his mind to starve the alien force out of his body, but gave into temptation and ate a bowl of sweet porridge that his wife Kasturba had prepared for him. 'This was sufficient invitation to the angel of death,' he wrote later. 'Within an hour the dysentery appeared in acute form.'[11]

It wasn't dysentery, but Spanish flu – a gastric variety in his case – and it incapacitated Gandhi at a critical moment in India's fight for independence. In 1918, he was forty-eight years old. He had returned to the land of his forefathers three years earlier, having spent two decades in South Africa learning the ropes of civil rights activism. Since then, his objectives had been twofold: to recruit for the British war effort, and to mobilise Indians by means of non-violent protest, or *satyagraha*. Some in the independence movement saw these goals as mutually exclusive. Not Gandhi. To him, India's contribution to the Allied war effort was a bargaining chip that could be exchanged, once the war had been won, for some degree of self-rule – dominion status, at the very least. *Satyagraha* was the stick to that carrot, a reminder to the British that Indians were prepared to fight peacefully for what was rightfully theirs.

Two of the earliest *satyagrahas* that Gandhi organised on Indian soil took place in Gujarat – the state in which he had been born, and where he built his ashram on his return from South Africa. The first, which got underway in February 1918, mobilised the textile workers of the largest Gujarati city, Ahmedabad, against low pay. A few months later, he persuaded the peasants of Kheda district, who had been pushed to the brink of starvation by the failure of the monsoon, to protest against the government's demands that they continue to pay a land tax.

Both *satyagrahas* had resulted in some if not all of the protesters' demands being met, and by the time he fell ill Gandhi was being seen in intellectual circles as a future leader of the nation. The trouble was, he lacked grass-roots support. In Kheda he had mobilised thousands, not hundreds of thousands. He considered it a start, the political awakening of the Gujarati peasant. But just how far he had to go was brought home to him that June, when he returned to Kheda to urge the peasants to enlist in the army. They refused. 'You are a votary of ahimsa [non-violence],' they pointed out. 'How can you ask us to take up arms?'

When the second wave of the Spanish flu erupted in September, it was exacerbated by drought. Water was desperately short that hot, dry autumn. 'People begged water,' one American missionary reported. 'They fought each other to get water; they stole water.'[12] In the countryside, cattle died for lack of grass, and bullocks had to be watched lest they leap into wells chasing the scent of damp. The first annual crop was due to be harvested, and the second sown, but with half the population sick there was not the manpower to complete these tasks. In Bombay Presidency, the province to which Gujarat belonged, staple food prices doubled. The government only halted the export of wheat in October, the month in which the epidemic peaked. By then, people were jumping moving freight trains to steal grain, and famished refugees were flooding into Bombay city, where cholera preyed on them too. Rivers became clogged with corpses because there wasn't enough wood to cremate them.

The colonial authorities now paid the price for their long indifference to indigenous health, since they were absolutely unequipped to deal with the disaster. In the presidency, their public health provision did not extend beyond the cities, and was anyway underpowered since many doctors were away at war. Nursing was in its infancy in India, and the only trained cohort of nurses was in Bombay city. Though more people were dying in cities than in rural areas, therefore, it was only in cities that help was to be had. Villages and remote communities were left, for the most part, to fend for themselves.

The government appealed for help, and it duly came – mostly from organisations with close links to the independence movement. Many of them were active in social reform, meaning they were well placed to mobilise dozens of local caste and community organisations. They raised funds and organised relief centres and the distribution of medicines, milk and blankets. In general, their efforts did not extend far beyond the urban centres either, but Gujarat was an exception. In that state, which is sometimes known as the cradle of free India – not only because Gandhi was born there, but also because of its long history of resisting colonial rule – something unusual happened.

While the municipal authorities in Ahmedabad were refusing to allow a school to be turned into a hospital (despite taxes having been raised to augment the salary of the city's health officer, something the local press was quick to point out), a pro-self-rule organisation that had helped organise the Kheda *satyagraha*, the Gujarat Sabha, set up an influenza relief committee to respond to the desperate need in the outlying villages of Ahmedabad district. Ambalal Sarabhai, one of the mill-owners who had held out against Gandhi's demands on behalf of the textile workers the previous February, even contributed funds to it.

Several hundred kilometres further south, in Surat district, freedom fighters also stepped into the breach – notably, three idealistic young men named Kalyanji and Kunvarji Mehta, who were brothers, and Dayalji Desai. They had been followers of the first leader of the independence movement, Bal Gangadhar Tilak, who was not against violence if it helped to achieve self-rule (Kunvarji Mehta had even assembled a bomb, though he had never detonated it). Over time, however, they had been won over by Gandhi's more peaceful methods. All three belonged to castes native to rural Gujarat – the Mehtas were middle-caste Patidars, Desai was a high-caste Anavil Brahmin – and during the 1910s, they had given up their government jobs to start ashrams in Surat. Their goal was to educate youngsters from their respective castes about India's struggle for freedom and the need for social reform – especially of the caste system.

The two ashrams – the Mehtas' and Desai's – now provided the manpower for a district-wide flu-relief operation. Funded by national pro-independence organisations, the Mehtas set up a free dispensary. Their students made the deliveries; Kalyanji himself went from house to house on a bike. They removed corpses for cremation. And when the Surat branch of Tilak's organisation, the Home Rule League, launched a vaccination programme, the two ashrams once again provided the volunteers. Their efforts were complemented by those of Surat's municipal commissioner who, more active than his counterpart in Ahmedabad, set up two travelling dispensaries and created an infectious-diseases ward at a local hospital.

What the student volunteers were vaccinating people with is not clear. Two vaccines were prepared that autumn, in two government laboratories, but they were made available in a very restricted way, and it wasn't until December – by which time the worst of the epidemic was over – that a new vaccine was distributed widely and free of charge. The medicines they dispensed were probably Ayurvedic. Western medicine wasn't yet widely accepted in India in 1918, and most people still relied on Ayurveda when ill. The efficacy of these medicines – like that of the vaccines – was questionable, but the students carried them out to the remotest villages in the district, and this brought them into contact with 'backward' social groups – including, for the first time, *adivasis*.

The *adivasis* (they would later be designated 'scheduled tribes') greeted the students with suspicion – these outsiders belonged to castes that had long exploited them – and many refused the medicines they brought. Some questioned the efficacy of Ayurveda, others argued that the only possible response to the sickness was to try to pacify the gods whose ire they had evidently provoked. Kalyanji Mehta's patience and pragmatism won them round, and many eventually took the medicine (his brother would go on to acquire the reputation of a miracle worker, for his efforts to improve their lives). The relief operation in Surat district is conservatively estimated to have reached 10,000 people – Hindus, Muslims, Christians, tribesmen and untouchables alike – and it earned the young freedom fighters the respect of the city-dwellers of Surat, Ahmedabad and Bombay, who read about their exploits in the papers.[13]

At Gandhi's ashram, meanwhile, several prominent members of the independence movement were laid low with flu that autumn, including the doughty widow Gangabehn Majmundar, the teacher of spinning on whom Gandhi was pinning his hopes for making India self-sufficient in cloth; his friend, the Anglican priest Charles Andrews; and Shankarlal Parikh, who had played a key role in the Kheda campaign. Gandhi was too feverish to speak or read; he couldn't shake a sense of doom: 'All interest in living had ceased.'

Medics came to give him the benefit of their advice, but he rejected most of it. Many of them remonstrated with him over his vow not to drink milk – the result of his disgust at the practice of *phooka*, in which air is blown forcefully into a cow's vagina to induce her to lactate. Supported by Kasturba, one doctor argued that on those grounds, he could have no objection to drinking goat's milk, since *phooka* was not practised on goats. He gave in, but bitterly regretted it later. To abandon one's guiding philosophy in the interests of living was unacceptable to Gandhi: 'This protracted and first long illness in my life thus afforded me a unique opportunity to examine my principles and to test them.' It's hard to say for sure, but his recovery may have been slow because he developed pneumonia. Before long, news of his illness – and of his obstinacy – spread, and the mouthpiece of the Gujarat Sabha, *Praja Bandhu*, castigated him: 'Mr Gandhi's life does not belong to him – it belongs to India.'

Gandhi was still unwell in November when he received the news that Germany had been defeated. The thought that he could now abandon his recruitment campaign came as a huge relief, but hardly had he begun to recover than he read in the papers about the publication of the Rowlatt Report. This was the incendiary document in which Justice Sidney Rowlatt of the viceroy's legislative council recommended the extension of martial law into peacetime in India. Throughout the war, civil liberties had been suspended, meaning that Indians could be arrested without charge and tried without a jury. Rowlatt found that levels of sedition and terrorism justified maintaining that situation. Indians had expected more freedom; instead they got more repression.

Rowlatt's bill passed into law in February 1919, triggering a wave of unrest. Gandhi was still weak. 'I could not at that time sufficiently raise my voice at meetings. The incapacity to address meetings standing still abides. My entire frame would shake, and heavy throbbing would start on an attempt to speak standing for any length of time.' But there was no question of him not rising to the occasion. To channel the disenchantment against what he called those 'black acts', he called for *satyagraha*. Dayalji Desai and Kalyanji Mehta answered his call in Surat. These two, whom caste barriers would normally have kept apart, were now united in the fight for self-rule under the nickname 'Dalu-Kalu'.

The *satyagraha* against the Rowlatt Act culminated in the tragic events of 13 April 1919, when Brigadier General Reginald Dyer ordered his troops to fire into an unarmed crowd in Amritsar, killing nearly 400 people, according to the government (more than 1,000 by other estimates). The British historian A. J. P. Taylor claimed that the Spanish flu led directly to this incident, by raising tensions in the country, and that it marked 'the decisive moment when Indians were alienated from British rule'.[14] Ten days later, an editorial appeared in the pro-independence *Young India* that reflected the nation's darkening mood. Entitled 'Public Health', it expressed the feeling on the streets of Bombay that a government that allowed 6 million to die of influenza (the contemporary estimate of the Indian death toll), 'like rats without succour', wouldn't mind if a few more died by the bullet. In May, just before he gave up his knighthood in protest over the Amritsar massacre, the poet Rabindranath Tagore wrote to a friend that the British were guilty of 'the same kind of ignorance of the eternal laws which primitive peoples show when they hunt for some so-called witch to whom they ascribe the cause of their illness while carrying the disease germs in their own blood'.[15]

In 1920, a special session of the Indian National Congress party was held in Calcutta. The Mehta brothers were among those who accompanied Gandhi in a special train from Bombay. When he promised self-rule within a year if Congress backed his call for nationwide *satyagraha*, Kunvarji Mehta was inspired. He returned to Gujarat and later delivered five towns to the cause. Half a million workers are estimated to have gone on strike in 1921, and many more in subsequent years. Gandhi's promise turned out to be premature; the bitter struggle for independence would drag on until 1947. But in 1921, thanks in no small part to the Spanish flu, he was the undisputed head of the independence movement, and he had grass-roots support.

21

Melancholy muse

Those who think about the Spanish flu at all often wonder why, when it left a cluster of tombstones in almost every graveyard in the world, it didn't weave a similar vein through the art of the time. The artists who attempted to depict it, to hitch up that train of zeroes to some notion of human suffering, are disconcertingly few. Why? It's a question that's received very little attention – one that's ripe for research. For now all we can do is sketch out the terrain and raise some hypotheses.

The first thing to be said is that art was not the same after the flu. The artistic waters didn't flow on smoothly, their surface untroubled. There was a rupture as violent as the parting of the Red Sea. All across the arts, the 1920s saw a desire to sever the link with Romanticism, to strip back, pare down, and slough off the exuberance of an earlier, misguided age. Painters and sculptors revisited classical themes. Architects jettisoned ornament and designed buildings that were functional. Fashion did something similar, dismissing colours and curves, while music underwent a number of parallel revolutions. Austrian composer Arnold Schönberg created a whole new musical system, dodecaphonics, while Russian-born composer Igor Stravinsky – under the influence of jazz – set out to replace feeling by rhythm.

It was a decade in which the artistic world turned its back on science and progress – a decade in which artists said, we had nothing on the ancients, after all. This new pessimism is usually attributed to the war. We're told it was humanity's response to death on an unimaginable scale. But there had been another, much bigger massacre against which all the achievements of science had proved helpless: the Spanish flu. It isn't possible to disentangle the effects of flu and war on the psyche of those who were alive then, but perhaps it isn't necessary. The challenge is more modest: to demonstrate that the Spanish flu contributed to that psychological shift.

The silence that is perhaps the most puzzling is the literary one. In his study of the flu in America, for example, Alfred Crosby noted that none of the 'supposedly hypersensitive' writers emerging just then in his country dealt with it – neither F. Scott Fitzgerald (who was caught by the tail end of it, while he was finishing off his first novel, *This Side of Paradise*), nor Ernest Hemingway (whose girlfriend, Agnes von Kurowsky, nursed flu-stricken soldiers in Italy), nor John Dos Passos (who came down with it on a military transport crossing the Atlantic), nor the doctor William Carlos Williams (who made sixty calls a day at the height of the crisis). Why did these writers ignore it?

To quote Maurois again, 'The minds of different generations are as impenetrable one by the other as are the monads of Leibniz.' Two things are worth noting, however. First, almost any writer you can name, who was adult in 1918, was touched either directly or indirectly by serious illness. Fitzgerald had TB, as did Anna Akhmatova and Katherine Mansfield; Hermann Hesse was turned down for military service in 1914, a dubious honour he shared with Hemingway; Tagore lost his wife and several of his children to disease; and both Luigi Pirandello and T. S. Eliot had wives who were considered insane. When Klimt waved a shotgun at the men who came to relieve him of *Medicine*, he was arguably 'speaking' for all of them.

Second, writers who were adult in 1918 had grown up in the Romantic tradition epitomised by Thomas Mann's *The Magic Mountain*, which he began writing in 1912 though it was not published until 1924. In this novel, the disease afflicting the residents of an alpine sanatorium represents Europe's moral decay on the eve of the First World War. For the Romantics, disease was symbolic – a metaphor for the sickness of the soul. It wasn't interesting in its own right, perhaps because they were bathed in it. It was too close to them; they couldn't see it. Things were changing, however. A year after *The Magic Mountain* came out, the British writer and regular invalid Virginia Woolf wrote an essay called *On Being Ill*, in which she asked why literature had not explored the rich terrain of illness: 'Considering how common illness is, how tremendous the spiritual change that it brings, how astonishing, when the lights of health go down, the undiscovered countries that are then disclosed . . . it becomes strange indeed that illness has not taken its place with love and battle and jealousy among the prime themes of literature.'

Her question could not be asked now, because starting in the 1920s disease moved centre-stage in literature – and no longer (or not only) as a symbol, but in all its ignominious, banal, terrifying reality. She herself contributed to that shift, exploring psychiatric illness in *Mrs Dalloway* (1925). *Ulysses* (1922) is peppered with allusions to bodily functions, and malfunctions, while in Eugene O'Neill's play *The Straw* (1919), which was inspired by his experiences in a TB sanatorium, disease doesn't stand for hell – it *is* hell. 'He sees life unsteadily and sees it black,' a critic wrote of O'Neill in 1921.[1]

What triggered this shift? Could it have been a virus that swept the globe in 1918, forcing infectious disease into human consciousness and highlighting the gap between the triumphant claims made for medicine and the dismal reality? The flu virus wasn't the only germ causing misery at the time. There were others – notably, the twin curses of TB and venereal disease – but these were chronic, slow-burners. They didn't come, cause devastation, and leave again, bringing in their wake a tsunami of lethargy and despair.

The Russian flu pandemic of the 1890s, it has been argued, contributed to the *fin de siècle* mood of cynicism and ennui.[2] It killed a million people; the Spanish flu killed at least fifty times as many. We don't know how many of the survivors suffered from post-viral fatigue, but the numbers must have been very large. And they were unlikely to have forgotten the puzzling randomness with which the flu had struck – that lethal lottery. Psychologists have an expression to describe the mindset of people subjected to random terror: learned helplessness. They tell us it leads to depression.

If you look hard, you can find traces of the Spanish flu in the writing of those who lived through it – heralds, perhaps, of the revolution to come. The disease left D. H. Lawrence with a weak heart and lungs, which he bequeathed to the gamekeeper, Mellors, in *Lady Chatterley's Lover* (1928). Katherine Anne Porter wrote *Pale Horse, Pale Rider* (1939) after catching the flu in Denver, Colorado at the age of twenty-eight (her black hair fell out, and when it grew back, it was white), while on the other side of the world, Saneatsu Mushanokōji – a member of the avant-garde Shirakaba or White Birch literary society in Japan – wrote a story about a young man returning from his travels in Europe, who learns that his girlfriend has died of influenza. Entitled *Love and Death* (1939) and still popular today, it describes a world full of happiness and light that suddenly turns black.

In September 1918, T. S. Eliot published a poem entitled 'Sweeney among the Nightingales', in which he makes a possible reference to the Spanish flu:

Gloomy Orion and the Dog
Are veiled; and hushed the shrunken seas;
The person in the Spanish cape
Tries to sit on Sweeney's knees

By November, the flu had interrupted normal life in almost every town and village in Britain. Both Eliot and his wife Vivien caught the disease, which apparently exacerbated Vivien's nervous condition to the extent that she found it impossible to sleep. She was living in Marlow, just outside London. He was in the capital itself, working on a vision of the desolate, haunted city that would become *The Waste Land* (1922) – itself possibly influenced by the strange atmosphere he imbibed at that time.

Intriguingly, post-viral fatigue leaves more of a trace than the flu itself, as if writers had mistaken the disease for the metaphor, and been tricked into giving it a proper treatment. One of the bestselling European novels of the 1920s, that caught the imagination of a generation, was Michael Arlen's *The Green Hat* (1924). Its protagonist, Iris Storm, is reckless, hedonistic and strangely detached from the world. She embodies many of the themes of the modern age: alienation, hypersensitivity, self-doubt. She was inspired by the heiress Nancy Cunard, who caught the flu in early 1919, developed pneumonia, and was dogged by depression throughout her long convalescence – the period in which Arlen knew her.

Another detached loner from the period is Sam Spade, the private detective in Dashiell Hammett's *The Maltese Falcon* (1929). Spade, the model for many later fictional detectives, finds a precursor in a little-known short story called 'Holiday' (1923), that the tubercular Hammett wrote after his own long and difficult recovery from the Spanish flu. It is about a tubercular soldier on day release from military hospital, a solitary man who lives only for the moment. In *The Maltese Falcon*, Spade recounts the parable of Flitcraft, a man who changes his life after nearly being killed by a falling beam: 'He knew then that men died at haphazard like that, and lived only while blind chance spared them.'

Modernism, which predated the war, provided the language that allowed artists and thinkers to explore the rich terrain that Woolf described. It liberated them from realism, from always being the outsider looking in, and in this it was influenced by psychoanalysis, which attached so much importance to dreams. Perhaps the lingering memory of those fever dreams contributed to this new fascination with the subconscious. The Polish composer Karol Szymanowski was staying in a Black Sea resort in the autumn of 1918, when he caught the Spanish flu and was inspired to write his opera *King Roger*. The 'Sicilian drama', as he called it, 'sprang into my mind one sleepless, Spanish night', after he and his cousin and librettist, Jarosław Iwaszkiewicz, had been strolling beside the azure sea. 'It seems to me,' Iwaszkiewicz wrote later, 'that this same intangible element of the eternal ocean, calming yet disquieting at the same time, became cast into the music which was subsequently composed.'[3] 'There was no light, there might never be light again, compared as it must always be with the light she had seen beside the blue sea that lay so tranquilly along the shore of her paradise,' wrote Porter in *Pale Horse*. 'There are better dreams' was Iris Storm's mantra.

But there was a newly black seam running through this exploration of the subconscious in the post-flu, post-war years. Sigmund Freud, the father of psychoanalysis himself, wrote an essay in 1920 entitled *Beyond the Pleasure Principle*, in which he introduced the concept of a death drive – *Todestrieb* – alongside that of the sex drive. At the time, he denied that the death from Spanish flu of his beloved daughter Sophie, pregnant with her third child, had influenced this development, but he later admitted that it may have played a role. 'Can you remember a time so full of death as this present one?' he wrote to his friend Ernest Jones, around the time of her demise, while to his widowed son-in-law, echoing Sam Spade, he wrote of 'a senseless, brutal act of fate'.[4] Psychoanalytic themes of sex and death permeated the first horror films, which were produced in the 1920s. *Nosferatu* (1922), directed by the German F. W. Murnau, was a retelling of the legend of Dracula, but with an additional subplot involving plague. The vampire is heading for Germany from his home in Transylvania, not far from the Black Sea, and spreading plague as he goes (ironically, the Spanish flu likely reached the Black Sea from Germany, brought by returning POWs).

Irony replaced pathos and, in the hands of writers like Pirandello (*Six Characters in Search of an Author*, 1921) and later Samuel Beckett (*Murphy*, 1938), tipped into absurdity. Kafka had long had an eye for the random and meaningless, and the Spanish flu must have struck him as a particularly fine example of the genre. 'Contracting fever as a subject in the Habsburg monarchy and re-emerging from it as a citizen of a Czech democracy was certainly eerie, though a bit comical, as well,'[5] wrote his biographer. When, having recovered, he stepped out into the streets of Prague, he discovered that they were full of people who had only just been enemies – French, Italians, Russians. There was no longer a Franz Joseph train station – it had been renamed Nádraži Wilsonovo (Wilson Train Station) – but there was now an October 28th Street, marking Czechoslovakia's birthday. And he wasn't the only one who felt as if he'd vanished down a rabbit hole. Both Gustav Landauer, a socialist who had been itching to take part in the revolution in Germany, and the fill-in chancellor, Max von Baden, woke from their respective fevers to find they had missed it. The philosopher and leading Zionist Martin Buber fell ill just as Europe's Jews looked to him for guidance in the matter of whether Palestine – which had recently passed from Ottoman to British control – could really be the homeland they had dreamed of.

Spanish writers and thinkers, whose identity had – against their will – become so tangled up with that of the flu, reacted to it in their own idiosyncratic way. Thanks to the operetta that had been on stage in Madrid when the spring wave had struck, and to deep-seated anxieties about the state of the nation, the disease had become inextricably linked in Spaniards' minds to Don Juan, the incorrigible libertine who, with all his strengths and weaknesses, stood in some way for all that was Spanish. Traditionally, All Saints' Day is marked in Spain by a performance of a version of the Don Juan myth, the play *Don Juan Tenorio*. By the time it came round in November 1918, however, Spaniards were in no mood for it. 'This year Don Juan has come at an inopportune moment,' wrote the critic José Escofet. 'We won't be able to attend'.[6]

After the pandemic, a number of Spanish writers set out either to parody the don, or to analyse and reform him. The philosopher Miguel de Unamuno was among them, as was his friend Gregorio Marañón, an eminent doctor and intellectual who had been involved in managing the disaster. A eugenicist like many of his contemporaries, Marañón believed that Spaniards were 'racially vigorous' but disadvantaged by their environment, in particular by the unhappy lot of women and children. In order that the stock should fulfil its potential, he felt, the cult of Don Juan had to be demolished, along with the implicit licence it gave to male promiscuity. In 1924 he wrote an essay pointing to the rake's lack of offspring, and suggesting that he may have been sterile, even effeminate. It was, arguably, the worst slur that could be levelled at one of the nineteenth century's great Romantic heroes.

More had died in the war than had died of flu, in Europe, but on every other continent, the opposite was true. If the pandemic contributed to a psychological shift in European literature, therefore, one might expect it to have done so to an even greater extent elsewhere. In Brazil, the departure of the Spanish flu marked a watershed moment. Doctors had been deeply unpopular in that country since Oswaldo Cruz had imposed a smallpox vaccination programme in 1904, but when cariocas saw that the flu was raging uncontrolled through Rio, they called for another well-known hygienist, Carlos Chagas – who was seen as Cruz's spiritual son – to step in. As soon as he did, the epidemic began, serendipitously, to recede, and from then on Brazilians looked at doctors with new respect.[7]

Brazil had been searching for a national identity ever since it had freed itself from its colonial masters in 1889, and doctors now gave it one. What defines a Brazilian, they said, is disease.[8] Disease – rather than race or climate – is the one thing that unites all social classes in Brazil. They talked about brazilianisation by infection, of Brazil as an immense hospital, and these ideas percolated into literature – reinforced, perhaps, by the memory of those flu-themed parades in the 1919 Rio Carnival, when groups calling themselves 'Midnight Tea' and 'Holy House' sang bawdy songs about a 'Spanish lady'.

In 1928, the writer Mário de Andrade published *Macunaíma*, a fable about a young man who was born in the Brazilian jungle with magical powers. Black, roguish, sensual, tricky, the eponymous Macunaíma represents the Brazilian personality, and he repeats the catchphrase 'Too little health and too many ants are the curses of Brazil.' Some writers were suspicious of the predominantly white doctors, however, seeing 'brazilianisation by infection' as thinly disguised eugenics. If Brazilians were sick, came their riposte, it was because of deep inequalities at the heart of Brazilian society. And so a literary counter-current emerged, that drew attention to those inequalities. Among those who contributed to it was the mixed-race writer Afonso Henriques de Lima Barreto, whose novella *Cemetery of the Living* (1956) compared the psychiatric hospital in which it was set to a cemetery, or hell.

The Spanish flu arrived in China at a time when the New Culture movement was challenging traditional Chinese values. It's hard to single out one epidemic from the many that battered that country at the time, but collectively, one could argue, they fuelled the drive to modernise. New Culture poured scorn on traditional Chinese medicine, which they saw as emblematic of all that was wrong with Chinese society, and they urged those in power to embrace western scientific ideas. One of the leaders of the movement was a little-known writer called Lu Xun. He had had his own scarring experience with Chinese doctors, having grown up with an ailing, alcoholic father. Each time the doctor called, he charged an exorbitant fee and sent Lu to gather the ingredients of a cure. They included a pair of crickets, the doctor having stipulated that 'They must be an original pair, from the same burrow.' Lu's father's health continued to deteriorate until he died, leaving his fourteen-year-old son to support the entire family.[9]

Lu studied western-style medicine in Japan, but later decided he could make a bigger difference with his pen. In 1919, he published a short story entitled 'Medicine', in which an elderly couple spend all their savings on a bread roll dipped in the blood of a recently executed criminal, believing that it will save their consumptive son – but he dies anyway: '"You there! Give me the money and you'll get the goods!" A man dressed in black stood before Shuan, who shrank back from his cutting glare. One enormous hand was thrust out, opened, before him; the other held, between finger and thumb, a crimson steamed bun, dripping red.'[10] Lu is now regarded as the father of modern Chinese literature.

Finally there was India, the country that had borne the brunt of the Spanish flu in terms of the sheer number of Indians who had died. Disease was a major preoccupation in the writing that emerged in that country in the 1920s, where it dovetailed with ideas about the need to reform the caste system and throw off the yoke of British rule. In China, the modernisers were campaigning for the replacement of literary language (*wenyan*) by spoken language (*pai-hua*) – the equivalent of replacing Latin with French or English during the European Renaissance – so that ordinary people could have access to Chinese culture. In India, something similar happened. The new generation of writers set out to describe the harsh realities of peasant life in language that, for the first time, peasants could understand. The most important of them was Munshi Premchand. Barely known on the global stage, unlike the Nobel Prize-winning Tagore, Premchand was arguably better loved in India. In *The Price of Milk* (1934), for example, he told the tale of Mangal, an untouchable orphan whose father had died of plague and his mother of snakebite. Mangal lives under a tree in front of his landlord's house, surviving on scraps. The landlord's wife won't touch him, for fear of pollution, even though Mangal's mother wet-nursed her son. The discrepancy requires no explanation because, as a priest remarks, 'Rajas and maharajas can eat what they want . . . Rules and restrictions are for ordinary people.'

Premchand became the self-styled 'chronicler of village life' around 1918, when he was living in the United Provinces (Uttar Pradesh), where the Spanish flu claimed an estimated 2–3 million lives alone. Also living there at that time was the poet Nirala, 'the strange one', who lost his wife and many other members of his family to the flu. He later recalled seeing the River Ganges 'swollen with dead bodies'. 'This was the strangest time in my life. My family disappeared in the blink of an eye.'[11]

These events, which took place when he was only twenty-two years old, left a deep impression on Nirala. A leading light in the Indian modernist movement, he had no patience with religious explanations of suffering that invoked karma or deeds done in previous lives. For him, the universe was a cruel place and there was no place in it for sentimentality. In 1921, he wrote a poem called 'Beggar', which arguably captured the mood, not only of Indian writers at that time, but of writers all over the world. It included the following lines:

> *When their lips shrivel up from starving*
> *what recompense*
> *from the generous Lord of destinies?*
> *Well, they can drink their tears.*

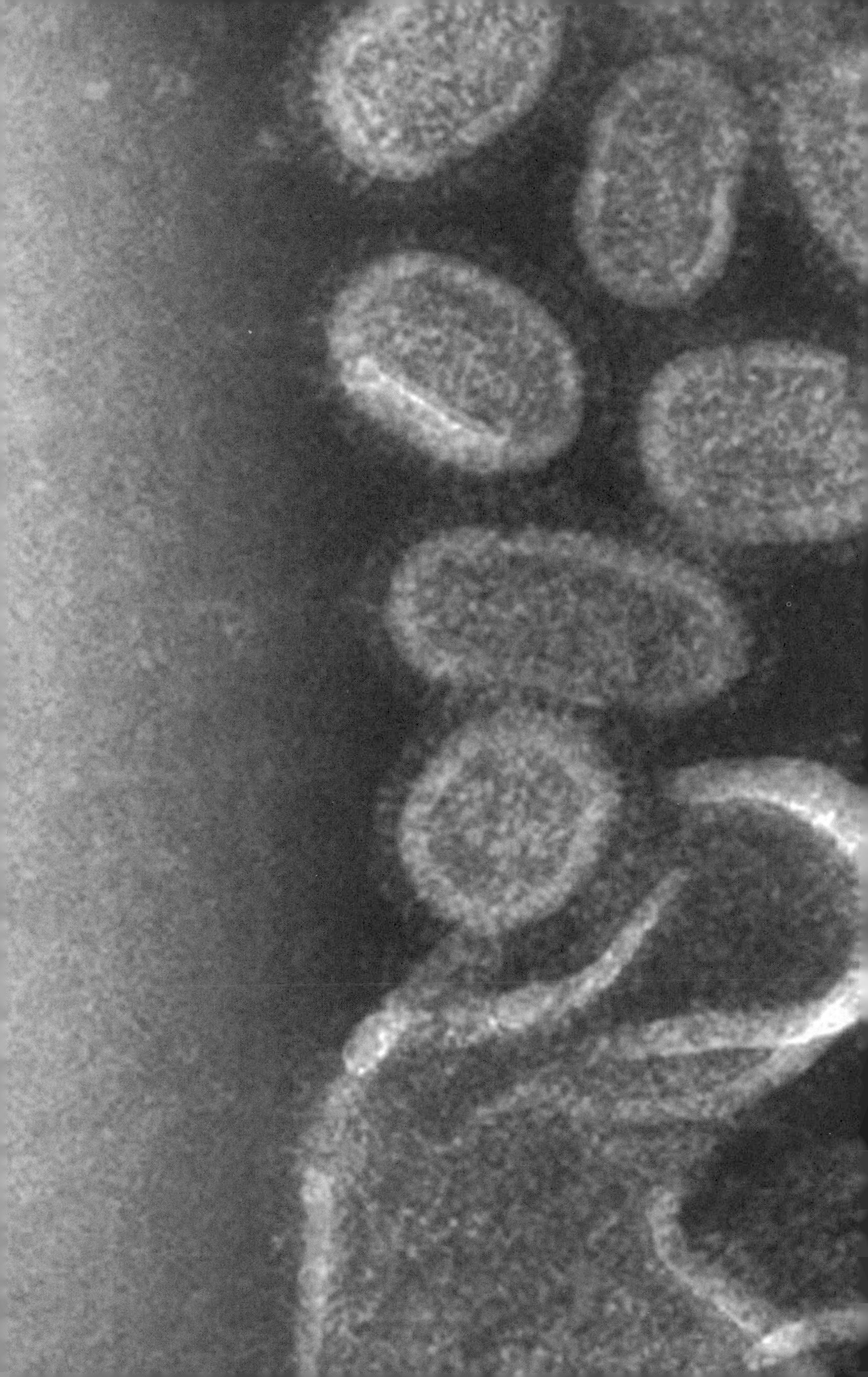

PART EIGHT: Roscoe's Legacy

Electron micrograph of the recreated 1918 flu virus

The 1995 film *Outbreak* tells the story of an outbreak of a fictional virus, Motaba, first in Zaire and then in a small town in America. Motaba resembles the real virus Ebola in that it causes a lethal haemorrhagic fever and – to begin with at least – it is transmitted by bodily fluids. At some point, however, it mutates and becomes airborne, like flu. To stop the virus spreading beyond the confines of the affected town, the president of the United States approves a plan to bomb that town. Thankfully, the plan is aborted.

This horrifying scenario has not come to pass. Ebola kills around half of those it infects, but it is not airborne, so it spreads much less easily than flu. The most vicious flu on record, on the other hand – the Spanish flu – killed 'only' a few per cent of those who caught it. The scientific consultants on *Outbreak* nevertheless insisted that the scenario was feasible. One of them was David Morens, the epidemiologist who, with Jeffery Taubenberger, dubbed the Spanish flu 'the mother of all pandemics'. He has even expressed the opinion that *Outbreak*'s scriptwriters could have gone further: 'I don't think they sensationalised it. If anything they toned it down.'[1]

A report published in 2016 by the Commission on Creating a Global Health Risk Framework for the Future (GHRF) – an independent, international group of experts convened by the US National Academy of Medicine – estimated there to be a 20 per cent chance of four or more pandemics occurring over the next hundred years, and a high probability that at least one of them will be flu.[2] Most experts consider it inevitable that there will be another flu pandemic. The only questions are when, how big, and what can we do to prepare ourselves? Lessons learned from the Spanish flu could help us to answer all three.

First let's address the issue of when. The Spanish flu occurred because a virulent viral strain acquired the ability, first to infect humans, and then to become highly transmissible between humans. It was this latter step that triggered the deadly autumn wave, and scientists now monitor strains in circulation in an attempt to predict when they might acquire that capability. One of the techniques they use is based, once again, on molecular clocks. The idea behind it is simple: as mutations accumulate over time, some of them may render a particular strain fitter or less fit than others. Those changes in fitness are reflected in the shape and branching of the flu family tree, because the fitter the strain, the more offspring it has. It is therefore possible, in theory, to predict when a particular strain might attain a level of fitness at which it has pandemic potential.

Strains may in fact already have emerged that have that potential. These belong to the H5N1 subtype of influenza A – the subtype that killed the toddler in Hong Kong in 1997. Almost all human cases of H5N1 infection to date have been caught directly from birds, but a few have been transmitted between humans, and some fear it is only a matter of time before the virus becomes highly transmissible between people (another strain, H7N9, is under surveillance for the same reason). That hasn't happened yet, and it may never happen, but since H5N1 also kills 60 per cent of those it infects, it is currently considered to rank among the world's greatest pandemic threats.

External factors – notably climate – might affect the timing of a pandemic. A study published in 2013 showed, for example, that prior to the Spanish flu and the three subsequent flu pandemics, the Pacific Ocean was in the La Niña phase of its temperature cycle.[3] During La Niña – known as the 'cold' phase of the El Niño-Southern Oscillation (ENSO) – the region of the Pacific between the tropics of Cancer and Capricorn cools, while in the opposite phase, El Niño, those same waters warm. Ocean and air currents are linked, since both redistribute heat around the earth's surface, and this has a knock-on effect on weather patterns around the globe, which is why meteorologists track ENSO so closely. (Could those portents seen prior to the Spanish flu – the withered roses, the owls turning up in new places – have been people's heightened perceptions of real atmospheric phenomena?)

El Niño ('little boy' in Spanish) episodes occur irregularly, but on average every two to seven years. They are sometimes, but not always, followed by La Niña ('little girl'). La Niña tends to last longer than El Niño, however – between one and three years, as opposed to less than a year for the little boy – and both tend to coincide with the northern hemisphere winter. Nobody yet knows why La Niña should make a pandemic more likely, but it may have something to do with the effect those changes in air currents have on the paths taken by migratory birds – and hence the populations of domesticated birds with which they come into contact.

Knowing that the world is about to enter a La Niña phase – as it did in August 2016 – could therefore help us to predict the next pandemic, though only as part of a bigger, more complex puzzle. If we understood the relationship between bird migration paths and flu, however, we might also be able to determine how our burning of fossil fuels will impact the timing, and geographical origin, of any future pandemic. We have, after all, now entered the Anthropocene epoch, which is defined by the impact of humanity on earth – the trace that our cars, nuclear weapons and discarded chicken bones have left on the planet. The previous epoch, the Holocene, spanned the 12,000 years since the last ice age and – coincidentally – the farming revolution that marked the beginning of the story of flu as a human disease. In the Anthropocene, we have moved into uncharted territory. As palaeo-climatologist William Ruddiman put it, 'we humans have now ended the 2.75-million-year history of northern hemisphere ice-age cycles for a time into the future that is beyond imagining'. In 2014, the Audubon Society of America found that birds had migrated further north by an average of sixty-four kilometres over the previous forty-eight years, as temperatures rose.[4] Global warming could even be having a direct effect on the flu virus itself – we don't know, but there are clues that this is the case. Cold, dry conditions tend to favour flu, but some strains currently in circulation appear to be adapting to a warmer world. There have been outbreaks of H5N1 in Asia in the summer months, for example.

So much for when. What about how big? This is the $64,000 question, because many different factors determine the scale of a flu pandemic. If the strain that caused the Spanish flu were to emerge again today, it would likely cause a mild disease, since our immune systems are more or less primed to it. The danger is that a new strain appears in our midst, to which nobody alive has ever been exposed. Even then, it's hard to predict what form a pandemic would take, because human beings have also moved on since 1918. The conditions that prevailed on the Western Front, and the massive displacement of people triggered by the First World War, are unlikely to be repeated. On the other hand, the globe is better connected. Transport, of humans and the germs that infect them, is faster, and we have fewer natural sanitary cordons in the form of geographic isolation. Our disease surveillance is better, and we have some effective medicines, including vaccines. But the world's population has also aged. Though age weakens the immune system, the elderly also have immune 'memories' of more varieties of flu, and it's not clear how those two effects would offset each other.

In 2013, a company that specialises in catastrophe modelling, AIR Worldwide, tried to take account of all these factors, and came up with an estimate of between 21–33 million deaths worldwide, if a flu as dangerous as the 1918 strain struck. The global population has roughly quadrupled since 1918, so this represents a very much smaller disaster than the Spanish flu, but it's still a staggering slew of death. It's also on the low side with respect to some of the other estimates that have been put forward over the years, which range from fewer than a million, to upwards of 100 million dead. Reflecting that huge span, there are those who say there is nothing to fear from a future pandemic, and others who lament how woefully underprepared we are. The former accuse the latter of being alarmist, the latter accuse the former of burying their heads in the sand. The chasm between them illustrates how much we still have to learn about pandemics in general, and about flu pandemics in particular.

Despite all the uncertainty, there *are* things we can do to prepare. The 2016 GHRF report called for governments and private and philanthropic bodies to stump up around $4 billion a year for pandemic preparedness, and it recommended that the money be invested in four main areas: a skilled and motivated public health workforce; robust disease surveillance systems; effective laboratory networks; and engagement with communities.

The Spanish flu and subsequent pandemics demonstrated that, given the right incentives and training, health workers stay at their posts and honour their duty to treat, often at great risk to their personal safety. That workforce therefore needs to be supported as much as possible and cared for in the event of illness. The best way to support them is to arm them with effective methods of surveillance and prophylaxis, and to make sure that they are dealing with an informed, compliant public. All three areas have seen huge advances since 1918, but there is still room for improvement.

At the moment, disease-surveillance agencies such as the CDC and the WHO take a good week to respond to a pandemic signal in the data. In 2009, two American researchers, Nicholas Christakis and James Fowler, set out to see if they could beat that, by identifying individuals who catch flu precociously in a pandemic, and who therefore act as 'sensors' of contagion. In an echo of Ronald Ross's 'theory of happenings', they realised that the way that anything contagious spreads through a population – be it a virus or a meme – depends on the structure of human social networks.

The key to their approach is something called the 'friendship paradox'. This is the idea that, on average, your friends have more friends than you have, and it arises because of a bias inherent in the way we count our friends (essentially, popular individuals get counted more often than less popular ones because they crop up in more people's social circles, so they inflate the average against which everyone compares themselves). For practical purposes, the friendship paradox means that if you pick a random person and ask them to nominate a friend, that friend is likely to be better connected than the person who nominated them. During the 2009 swine flu outbreak, Christakis and Fowler tracked infection in two groups – one randomly chosen group of Harvard undergraduates, and a second group whom the first had nominated as friends. They found that the friends fell sick on average *two weeks* earlier than their randomly picked counterparts – presumably because they were more likely to come into contact with carriers of infection.[5]

If you could capture that early spike in flu cases, and mobilise a containment strategy up to two weeks earlier than is possible now, you could potentially save a great many extra lives. A lot of vulnerable people can be vaccinated in two weeks. But there is another way in which those sensors could help limit the impact of a pandemic – or even avert it entirely. If a high enough proportion of a population is vaccinated prior to a pandemic, they may confer what is known as 'herd immunity' on the rest. This is because they block the virus's spread, meaning that the whole population is protected even though not everyone is immune. Christakis and Fowler showed that herd immunity could be achieved by vaccinating a smaller number of sensors than less well-connected people – again, because they were more likely to step into the paths of carriers of infection.

What about prophylaxis? The annual flu vaccine is improving all the time, but it still has to be updated each year. Since 1973, the WHO has issued annual recommendations regarding which strains should go into it, depending on those that surveillance agencies indicate are currently circulating in the human population. It takes time to manufacture a new flu vaccine, however, so final decisions about the composition tend to be taken around February for a vaccination period that begins in October. And therein lies a problem: if a new strain surfaces between February and October, the vaccine will be partially effective at best. The molecular clocks might help prevent that happening, too, by making it possible to identify strains whose fitness is increasing, though they haven't yet been detected as a threat.

Meanwhile, work continues on a so-called 'universal' vaccine – one that will protect humans against flu *without* having to be updated each year. For some time now, vaccines have not made use of the entire flu virus, because exposure to this in the form of a vaccine can cause side effects that are sometimes more unpleasant than the flu itself. To stimulate a response from the human immune system, modern vaccines present it with the round, convoluted head of the H antigen alone. Unfortunately, it is precisely this that changes from year to year, so Taubenberger for one is pursuing an alternative approach.

During his studies of the Spanish flu, he realised that another part of the H antigen does not change from year to year: the stem. That's because it has to anchor the head in place, meaning it is subject to certain mechanical constraints. His group, among others, is now focusing on this essential but relatively unchanging component of the virus, in an attempt to develop a vaccine that protects, not only against all the flu strains that have caused pandemics in the past, but potentially also against those that could do so in future.

In a future flu pandemic, health authorities will introduce containment measures such as quarantine, school closures and prohibitions on mass gatherings. These will be for our collective benefit, so how do we ensure that everyone complies? How, too, do we persuade people to get vaccinated each year, given that herd immunity is the best protection we have against a flu pandemic? Experience has shown that people have a low tolerance for mandatory health measures, and that such measures are most effective when they are voluntary, when they respect and depend on individual choice, and when they avoid the use of police powers. In 2007, the CDC issued guidelines for how to ensure maximum compliance with public health measures in a pandemic. Based partly on lessons learned in 1918, these recommended that measures only be made mandatory when the proportion of the sick who die rises above 1 per cent (remember that this proportion was at least 2.5 per cent for the Spanish flu). Using 2016 numbers, that means that more than 3 million Americans would have to die before the CDC would advise such a step – a measure of how counterproductive that organisation believes compulsion to be.

But if disease containment works best when people choose freely to comply, then people must be informed about the nature of the disease and the risk it poses. This is one reason why it's important to tell the story of the Spanish flu. It's also one of the arguments used to justify films such as *Outbreak*. Presenting the worst-case scenario, the defenders of such films claim, is the best way to persuade people to get vaccinated and to keep funding scientific research via their taxes and private donations. It's a controversial strategy, however, not only because of the danger that such films will provoke 'apocalypse fatigue', but also because the ability of scientists to predict the worst-case scenario depends on how well they understand the phenomenon in question. H5N1 might still turn out to be as dangerous as the fictional Motaba – we'll have to wait and see. But in the early twentieth century, eugenics-inspired movies frightened people by showing them the supposed implications of their ill-advised reproductive choices for society – to wit, the proliferation of 'defectives' – and eugenics has since been thoroughly discredited.

Whatever the rights and wrongs of such shock tactics, the media clearly have a critical role to play in any future pandemic, and 1918 taught us a valuable lesson in this too: censorship and playing down the danger don't work; relaying accurate information in an objective and timely fashion does. Information and engagement are not the same thing, however. Even when people have the information they need to contain the disease, they do not necessarily act on it. A few years ago, when the European Commission ordered the destruction of olive trees in the Italian region of Puglia, to prevent the spread of a dangerous plant pathogen, local people protested and challenged the decision in the courts. Olive trees have deep emotional significance in Italy, where families plant them to mark births over generations. The EC had not involved the olive-tree owners in their deliberations, and the owners rejected the scientific arguments it advanced.[6] Trust broke down between the two parties – or rather, was never built up. But trust is not something that can be built up quickly. If it is not in place when a pandemic declares itself, then however good the information being circulated, it probably won't be heeded.

Another thing 1918 taught is that, sometimes, the reasons why people ignore advice are to be found deep in the past. In this century, President Thabo Mbeki of South Africa denied that AIDS was caused by a virus and appointed a health minister who recommended treating it with garlic, beetroot and lemon juice. Soon enough, AIDS patients were dying on the front lawns of hospitals because they were unable to get effective treatment. Mbeki's behaviour seems impossible to understand, until you set it in the context of a long history of whites blaming blacks for disease in his country. The consequences of that blame have often been brutal and long-lasting for black South Africans, as they were in 1918. The pandemic forced action on an issue that had been under discussion for a decade: segregation of towns along colour lines. In 1923, the Natives (Urban Areas) Act was passed, and it wouldn't be repealed for another sixty years.

Through such secondary tragedies the Spanish flu cast a long shadow over humanity. Some of those tragedies could not have been avoided by good disease surveillance or a vaccine, but others could – the surge in post-viral depression, the creation of large numbers of orphans, the damaging of the life chances of the generation in the womb. That so much suffering is now preventable is a testament to the fact that Roscoe Vaughan, an anonymous woman in an Alaskan grave, and others whose tissues allowed Taubenberger and Reid to sequence the flu genome, did not lose their lives in vain. But we shouldn't rest on our laurels, because the story doesn't end here. Once people thought that flu was caused by the pull of distant stars. Then they realised that something very small penetrated the body and made it sick. Finally, they understood that influenza is the product of an interaction between a host and an agent of disease. Over the centuries, humans came to perceive flu as an increasingly intimate dance with the Devil, and even as they add to their knowledge, man and microbe continue to shape one another.

AFTERWORD: On Memory

Sincerely Yours, Woodrow Wilson. *Arthur Mole designed the portrait and photographed the scene after his partner, John Thomas, arranged the 21,000 soldiers on parade grounds at the US Army's Camp Sherman, 1918.*

> Whenever anyone asked Samuel about his parents, he would tell them that they had died of the Spanish flu. And if someone replied that this was quite impossible, since the Spanish flu epidemic had reached Brazil at the beginning of the twentieth century, he would respond: 'Well, maybe it was Asian flu; I didn't ask it for its passport.'
>
> Elias Canetti, *Party in the Blitz*

Arthur Mole was a man of unusual vision. During the First World War, armed with a white flag and a megaphone, he choreographed tens of thousands of US soldiers into what he called 'living photographs'. If you looked at the mass of men from the ground, or from directly above, they looked like a mass of men. But if you stood at the top of a twenty-five-metre-high viewing tower placed a certain distance away, you saw that they formed a patriotic image: the Statue of Liberty, Uncle Sam, the head of President Wilson.

Mole understood that meaning comes with distance. The Spanish flu has been called the forgotten pandemic, but it isn't forgotten. Our collective memory of it is simply a work in progress. A hundred years on, we have some distance on it, though Mole would probably carry on marching before he put down his viewing platform – until he reached the point at which the lines of perspective converge to produce an intelligible image.

The lines of perspective never really converge, when it comes to history, so off Mole marches on his eternal quest, until he vanishes. Looking back the other way, to a distance of just under 700 years, we see the Black Death hove into view. In the mid-fourteenth century, the worst pandemic in human history killed an estimated 50 million people, though the numbers are even sketchier than for the Spanish flu, and may in reality have been far higher. The Black Death is certainly not forgotten, nor is it overshadowed in our minds by a war with which it coincided – the Hundred Years War – yet our collective memory of it took time to coalesce. Even in 1969, Philip Ziegler, the author of an excellent account of that plague, was able to write that 'There are remarkably few full-length studies dealing with the Black Death as a whole or even in a country or group of countries.' Some may have been lost, but of the six surviving studies that he considered most important, the earliest was published in 1853, 500 years after the event. The pandemic did not even acquire the name by which we know it until the sixteenth century. In the Middle Ages, it was known as the 'Blue Death'.

Wars and plagues are remembered differently. Collective memories for war seem to be born instantly, fully formed – though subject, of course, to endless embellishment and massage – and then to fade over time. Memories of cataclysmic pestilence build up more slowly, and once they have stabilised at some kind of equilibrium – determined, perhaps, by the scale of death involved – they are, in general, more resistant to erosion. The sixth-century Plague of Justinian is remembered better today than the eighth-century An Lushan Rebellion in China, though to the best of our knowledge, they killed comparable numbers.

We find ourselves at an interesting point on the remembering/forgetting arc with respect to the twentieth century. The two world wars are still raw, we refer back to them obsessively and are firmly convinced that we will never forget them – though past experience suggests that they will gradually lose their lustre in our minds, or be obscured by other wars. Meanwhile the Spanish flu intrudes more and more into our historical consciousness, but it can't shake the prefix 'forgotten'.

WorldCat, the world's largest library catalogue, currently lists around 80,000 books on the First World War (in more than forty languages), and around 400 on the Spanish flu (in five languages). But those 400 books represent an exponential increase over what had been written on the subject twenty years ago. The range of academics who show an interest in it is now very broad, and it isn't just academics. In the twenty-first century – a century in which writers have firmly embraced illness as a subject worthy of treatment, alongside love, jealousy and battle – the Spanish flu has finally penetrated popular culture, providing plot lines for novels, movies and TV dramas.[1] In the popular British TV series *Downton Abbey*, for example, three main characters catch Spanish flu in April 1919, and one dies of it. In 1921, American sociologist James Thompson compared the fallout of the Black Death with that of the First World War.[2] The Spanish flu would arguably have been a more natural reference for comparison with the other plague, and yet only two years after it, it didn't occur to him to use it. Nor did the Spanish flu register on Ziegler's historical radar when he cited Thompson's paper nearly fifty years later. Such an oversight would not be possible today.

Why does memory for a pandemic take time to develop? Perhaps one reason is that it's not so easy to count the dead. They don't wear uniforms, display exit wounds or fall down in a circumscribed arena. They die in large numbers in a short space of time, over a vast expanse of space, and many of them disappear into mass graves, not only before their disease has been diagnosed, but often before their lives have even been recorded. For most of the twentieth century, people thought the Spanish flu had killed around 20 million people, when the real number was two, three, possibly even five times that.

And then, the Spanish flu is a difficult pandemic to pigeonhole. It killed horribly, and it killed many more of its victims than any other flu pandemic we know of, yet for around 90 per cent of those who caught it the experience was no worse than a dose of seasonal flu. As a result, people didn't know how to think about it; they still don't. At the time many mistook it for pneumonic plague – a disease that can be transmitted directly between people and that, unless treated, is almost always lethal. Today, they shudder at the thought of the hypothetical equivalent – airborne Ebola. But in general, the Spanish flu was far more mundane than either of those things. It was also very rapid, burning itself out in any given place before a siege mentality could set in. Epidemics of bubonic plague and AIDS, in contrast, linger in one locality for years.

Memory is an active process. Details have to be rehearsed to be retained, but who wants to rehearse the details of a pandemic? A war has a victor (and to him, the spoils – the version that is handed down to posterity), but a pandemic has only vanquished. Until the nineteenth century, pandemics were considered acts of God and people accepted them fatalistically, but with the advent of germ theory, scientists realised that they could, in principle, prevent them. Their inability to do so in 1918 was humiliating, a reminder of earlier epochs when epidemics came without rhyme or reason, and there was nothing they could do to stop them. As one epidemiologist put it, 'It was as if one of the old plagues had returned.'[3]

At least one constituency, therefore – and a powerful one at that – had reason to pass over the Spanish flu in silence. The philosopher Walter Benjamin even argued that such public silences are essential to progress, because they allow us to leave behind the ruins of the past. So the Yupik of Bristol Bay made a pact – *nallunguaq* – not to speak about the pandemic that had shattered their ancient culture. When the story of the Spanish flu was told, it was told by those who got off most lightly: the white and well off. With very few exceptions, the ones who bore the brunt of it, those living in ghettoes or at the rim, have yet to tell their tale. Some, such as the minorities whose languages died with them, never will. But perhaps the victims did find a way to express themselves, after all – in strikes, protest and revolution.

There is another reason why a pandemic memory might take time to mature. In 2015, psychologists Henry Roediger and Magdalena Abel of Washington University in St Louis, Missouri summed up a still thin body of research into collective memory when they wrote that its narrative structure 'is rather simple and comprises only a small number of salient events referring to beginning, turning, and end points'.[4] It helps, they added, if those events have heroic or mythical components. Wars slot easily into that structure, with their declarations and truces, their acts of outstanding bravery. A flu pandemic, on the other hand, has no clear beginning or end, and no obvious heroes. The French war ministry tried to create some, by awarding a special 'epidemic medal' to thousands of civilian and military personnel who had shown devotion in the fight against the disease, but it didn't work. A war-memorabilia website notes that 'Curiously, its place among the important decorations of that conflict is completely unknown.'[5]

A different narrative structure is needed, and a new language. Piqued by their humiliation, scientists went on to furnish us with a vocabulary of flu – with concepts such as immune memory, genetic susceptibility and post-viral fatigue. Couched in this new language – not a poetic language, perhaps, but one that allowed you to make predictions, and to test them against the historical reports – disparate events began to appear connected, while other, once obvious links atrophied and died (no, it wasn't the punishment of an angry god; yes, it was at least partly responsible for the subsequent wave of melancholy). The pandemic took on a radically new shape: the one we recognise today.

Such a narrative takes time to develop – around a hundred years, judging by the burst of interest in the last two decades – and until it does, all kinds of confusion arise. In Australia, the Spanish flu became telescoped in people's minds with a 1900 outbreak of bubonic plague, in part because newspapers referred to both as 'plague', while in Japan it was eclipsed by another natural disaster, the Great Kantō earthquake of 1923, which destroyed Tokyo. Many people thought the flu was the product of biowarfare, and the flu and the war were conflated or confounded in other ways too. The captains and lieutenants who died while serving with the British Army – Vera Brittain's 'lost generation' – numbered around 35,000.[6] But six times as many Britons died of Spanish flu, and half of those were in the prime of life – young, fit men and women whose promise also lay ahead of them. They may therefore be considered more deserving of the label 'lost generation', though the flu orphans, and those who were in their mother's womb in the autumn of 1918, may lay claim to it too, for different reasons.

The death of Edmond Rostand illustrates this imaginative fusing of war and flu. On 10 November 1918, he was getting ready to leave his home in the Basque Country to celebrate the imminent armistice in Paris. At five in the afternoon the car arrived that was to take him and his mistress, Mary Marquet, to the station. While their luggage was being loaded, the pair sat close to the fire, watching the dying embers. They were wistful, even sombre: a dangerous disease was raging in Paris, major events were unfolding on the world stage. Suddenly they heard wings fluttering against a window. Rostand went to open it and a dove entered, staggering towards the hearth. He bent to pick it up, but as he cupped it in his hands, its wings grew slack. 'Dead!' he cried. Marquet, shocked, murmured that it was a bad omen, and three weeks later the celebrated creator of Cyrano de Bergerac died in Paris of the Spanish flu.[7] It would be hard to think of a more fitting symbol of the twin dangers hanging over humanity at that moment than a sick dove.

The 1918 pandemic is still emerging from the shadows of the First World War, but emerge it will, because of what we have already come to understand about it. The Spanish flu was an example of what today we would call, with appropriate avian overtones, a black-swan event. No European thought that black swans existed until a Dutch explorer discovered them in Australia in 1679, but as soon as he had, all Europeans realised that black swans had to exist, since other animals came in different colours. Likewise, though there had never been a flu pandemic like 1918 before, once 1918 had happened, scientists realised that it could happen again. Hence the reconstructed virus is maintained in high containment facilities where scientists study it in the hope of developing better vaccines; art historians pore over the self-portraits of famous survivors looking for traces of post-viral fatigue; and novelists try to put themselves inside the heads of those who lived through it, in an attempt to understand their fear. Like worker bees they are busy weaving threads between the millions of discrete tragedies to create a collective memory – a living photograph of the Spanish flu. Those threads will reinforce it in our consciousness, and help pull it free.

Acknowledgements

For the idea, thanks to Richard Frackowiak. For their invaluable help with the research and in some cases the translation, thanks to Robert Alexander, Séverine Allimann, Andrey Anin, Pierre Baudelicque, Annette Becker, Charles Linus Black, Elizabeth Brown, Ivana Bucalina, Marta Cerezo Guiu, Upendra Dave, Jean-René and François Dujarric de la Rivière, Mark Elgar, Thomas Fischer, Sofie Frackowiak, Paul French, John Garth, Douglas Gill, Anders Hallberg, Claude Hannoun, Jean-Frédéric Henchoz, Laura Jambrina, Peter Johnson, Andreas Jung, Bahri Karaçay, Ana Leal, Ahnie Litecky, Daniel Medin, Jürgen Müller, Sandy Rich, Inna Rikun, Nil Sari, Janice Shull, Maria Sistrom, Stéphanie Solinas, Tim Troll, Malvina Vlodova, Liliya Vukovich and the staff of the Gorky Research Library in Odessa, Jeanine Wine, Negar and Mohammad Yahaghi, and Patrick Zylberman. For their editorial advice, which made it a better book, thanks to Alex Bowler, Ana Fletcher, Janet Lizop, Michal Shavit and Jeffrey Taubenberger. For their endless hospitality, thanks to Pamela and Gian Luigi Lenzi. For their grant, without which much of the research would have been impossible, thanks to the Society of Authors' K. Blundell Trust and Authors' Foundation. And for his wisdom, wit and kindness, a heartfelt tribute to David Miller, who died too soon and before this book saw the light of day.

Illustration Credits

The images used to introduce each of the ten sections of this book have been reproduced with kind permission from the following:

Introduction: Bettmann / Contributor / Getty Images / Editorial #: 514877124. Part One: Popperfoto / Contributor / Getty Images / Editorial #: 79035213. Part Two: OHA 250: New Contributed Photographs Collection, Otis Historical Archives, National Museum of Health and Medicine (NCP 001603). Part Three: The Family, 1918, Schiele, Egon (1890–1918) / Osterreichische Galerie Belvedere, Vienna, Austria / Bridgeman Images. Part Four: Harris & Ewing collection, Prints & Photographs Division, Library of Congress, LC-DIG-hec-44028. Part Five: © IWM (Q 2381). Part Six: Institut Pasteur/Musée Pasteur. Part Seven: Charles Linus Black; Photographer: Sue Brown French. Part Eight: Photo credit: Cynthia Goldsmith; content provider(s) CDC / Dr Terrence Tumpey. Afterword: Bettmann / Contributor / Getty Images / Editorial #: 514865716.

Notes

INTRODUCTION: THE ELEPHANT IN THE ROOM

1. The exact quote is: 'The brevity of the influenza pandemic of August–September 1918 posed great problems to doctors at the time. They had no chance to try out different remedies or to learn anything about the disease before it was over. It has posed great problems to historians ever since.' T. Ranger, 'A historian's foreword', in H. Phillips and D. Killingray (eds.), *The Spanish Influenza Pandemic of 1918–19: New Perspectives* (New York: Routledge, 2003), pp. xx–xxi. This is also the source for Ranger's comments on the need for a new style of narrative.
2. J. Winter, *Sites of Memory, Sites of Mourning: The Great War in European Cultural History* (Cambridge: Cambridge University Press, 1995), p. 20.
3. Ranger, in Phillips and Killingray (eds.), pp. xx–xxi. Ranger is thinking particularly of the female characters in the novels of Zimbabwean writer Yvonne Vera, whose surname was one of the names given to the Spanish flu in her country.
4. L. Spinney, 'History lessons', *New Scientist*, 15 October 2016, pp. 38–41.

PART ONE: THE UNWALLED CITY

1. Coughs and sneezes

1. N. D. Wolfe, C. P. Dunavan and J. Diamond, 'Origins of major human infectious diseases', *Nature*, 17 May 2007; 447(7142):279–83.
2. Epicurus, *Vatican Sayings*.
3. Book 25, *The Fall of Syracuse*.
4. W. H. McNeill, *Plagues and Peoples* (Garden City: Anchor Press/Doubleday, 1976), p. 2.
5. D. Killingray, 'A new "Imperial Disease": the influenza pandemic of 1918–19 and its impact on the British Empire', paper for the annual conference of the Society for Social History of Medicine, Oxford, 1996.

6. W. F. Ruddiman, *Earth Transformed* (New York: W. H. Freeman, 2013), ch. 21.
7. C. W. Potter, 'A history of influenza', *Journal of Applied Microbiology* (2001), 91:572–9.
8. *Quick Facts: Munch's The Scream* (Art Institute of Chicago, 2013), http://www.artic.edu/aic/collections/exhibitions/Munch/resource/171.

2. The monads of Leibniz

1. P. de Kruif, *Microbe Hunters* (New York: Harcourt, Brace & Co., 1926), pp. 232–3.
2. *Ulysses*, 2:332–7.
3. Hippocrates. *Ancient Medicine*.
4. T. M. Daniel, 'The history of tuberculosis', *Respiratory Medicine*, 2006; 100:1862–70.
5. S. Otsubo and J. R. Bartholomew, 'Eugenics in Japan: some ironies of modernity, 1883–1945', *Science in Context*, Autumn–Winter 1998; 11(3–4):545–65.
6. G. D. Shanks, M. Waller and M. Smallman-Raynor, 'Spatiotemporal patterns of pandemic influenza-related deaths in Allied naval forces during 1918', *Epidemiology & Infection*, October 2013; 141(10):2205–12.
7. J. Black and D. Black, 'Plague in East Suffolk 1906–1918', *Journal of the Royal Society of Medicine*, 2000; 93:540–3.
8. A. D. Lanie et al., 'Exploring the public understanding of basic genetic concepts', *Journal of Genetic Counseling*, August 2004; 13(4):305–320.

PART TWO: ANATOMY OF A PANDEMIC

3. Ripples on a pond

1. B. Echeverri, 'Spanish influenza seen from Spain', in Phillips and Killingray (eds.), p. 173.
2. E. F. Willis, *Herbert Hoover And The Russian Prisoners Of World War I: A Study In Diplomacy And Relief, 1918–1919* (Whitefish: Literary Licensing, LLC, 2011), p. 12.
3. D. K. Patterson and G. F. Pyle, 'The geography and mortality of the 1918 influenza pandemic', *Bulletin of the History of Medicine*, Spring 1991; 65(1):4–21.

4. R. Hayman, *A Life of Jung* (London: Bloomsbury, 1999). But Hayman gives no source for this story, and according to Thomas Fischer, director of the Foundation of the Works of C. G. Jung, there is no documentary evidence for it.
5. E. Favre, *L'Internement en Suisse des Prisonniers de Guerre Malades ou Blessés 1918–1919: Troisième Rapport* (Berne: Bureau du Service de l'Internement, 1919), p. 146.
6. *My Life and Ethiopia's Progress, 1892–1937: The Autobiography of Emperor Haile Selassie I*, ed. E. Ullendorff (Oxford: Oxford University Press, 1976), p. 59.
7. R. Buckle, *Diaghilev: biographie*, translated by Tony Mayer (Paris: J-C Lattès, 1980), p. 411.
8. R. Stach, *Kafka: The Years of Insight*, translated by Shelley Frisch (Princeton: Princeton University Press, 2013), pp. 252–5.
9. S. Słomczyński, '"There are sick people everywhere – in cities, towns and villages": the course of the Spanish flu epidemic in Poland', *Roczniki Dziejów Społecznych i Gospodarczych*, Tom LXXII – 2012, pp. 73–93.
10. A. W. Crosby, *America's Forgotten Pandemic: The Influenza of 1918* (Cambridge: Cambridge University Press, 1989), p. 145–50.
11. French Consul General's report on sanitary conditions in Milan, 6 December 1918, Centre de documentation du Musée du Service de santé des armées, Carton 813.
12. R. F. Foster, *W. B. Yeats: A Life, Volume II: The Arch-Poet 1915–1939* (New York: Oxford University Press, 2003), p. 135.
13. W. Lanouette, *Genius in the Shadows: A Biography of Leo Szilard* (New York: Charles Scribner's Sons, 1992), pp. 41–2.
14. H. Carpenter, *A Serious Character: the Life of Ezra Pound* (London: Faber & Faber, 1988), p. 337.
15. G. Chowell et al., 'The 1918–1920 influenza pandemic in Peru', *Vaccine*, 22 July 2011; 29(S2):B21–6.
16. A. Hayami, *The Influenza Pandemic in Japan, 1918–1920: The First World War between Humankind and a Virus*, translated by Lynne E. Riggs and Manabu Takechi (Kyoto: International Research Center for Japanese Studies, 2015), p. 175.

4. Like a thief in the night

1. N. R. Grist, 'Pandemic influenza 1918', *British Medical Journal*, 22–9 December 1979; 2(6205):1632–3.
2. N. P. A. S. Johnson, *Britain and the 1918–19 Influenza Pandemic: A Dark Epilogue* (London: Routledge, 2006), pp. 68–9.
3. L. Campa, *Guillaume Apollinaire* (Paris: Éditions Gallimard, 2013), p. 764.
4. Letter written to Richard Collier by Margarethe Kühn, 26 April 1972. Unpublished. In the collection of the Imperial War Museum, London.
5. J. T. Cushing and A. F. Stone (eds.), *Vermont in the world war: 1917–1919* (Burlington, VT: Free Press Printing Company, 1928), p. 6.
6. C. Ammon, 'Chroniques d'une épidémie: Grippe espagnole à Genève', PhD thesis (University of Geneva, 2000), p. 37.
7. M. Honigsbaum, *Living with Enza: The Forgotten Story of Britain and the Great Flu Pandemic of 1918* (London: Macmillan, 2009), p. 81.
8. The title of Porter's story, from which this book takes its own title, references an African-American spiritual, which in turn references Revelations 6:8: 'And there, as I looked, was another horse, sickly pale; and its rider's name was Death, and Hades came close behind. To him was given power over a quarter of the earth, with the right to kill by sword and by famine, by pestilence and wild beasts.'
9. M. Ramanna, 'Coping with the influenza pandemic: the Bombay experience', in Phillips and Killingray (eds.), p. 88.
10. P. Nava, *Chão de ferro* (Rio de Janeiro: José Olympio, 1976), ch. 2: 'Rua Major Ávila'.
11. S. C. Adamo, 'The broken promise: race, health, and justice in Rio de Janeiro, 1890–1940', PhD thesis (University of New Mexico, 1983), p. iv.
12. H. C. Adams, 'Rio de Janeiro – in the land of lure', *The National Geographic Magazine*, September 1920: 38(3):165–210.
13. T. Meade, *'Civilising' Rio: Reform and Resistance in a Brazilian City, 1889–1930* (University Park: Penn State University Press, 1996).
14. A. da C. Goulart, 'Revisiting the Spanish flu: the 1918 influenza pandemic in Rio de Janeiro', *História, Ciências, Saúde – Manguinhos*, January–April 2005; 12(1):1–41.
15. Ibid.
16. R. A. dos Santos, 'Carnival, the plague and the Spanish flu', *História, Ciências, Saúde – Manguinhos*, January–March 2006; 13(1):129–58.

PART THREE: *MANHU*, OR WHAT IS IT?

5. Disease eleven

1. World Health Organization Best Practices for the Naming of New Human Infectious Diseases (Geneva: World Health Organization, May 2015), http://apps.who.int/iris/bitstream/10665/163636/1/WHO_HSE_FOS_15.1_eng.pdf?ua=1
2. R. A. Davis, *The Spanish Flu: Narrative and Cultural Identity in Spain, 1918* (New York: Palgrave Macmillan US, 2013).
3. J. D. Müller, 'What's in a name: Spanish influenza in sub-Saharan Africa and what local names say about the perception of this pandemic', paper presented at 'The Spanish Flu 1918–1998: reflections on the influenza pandemic of 1918–1919 after 80 years' (international conference, Cape Town, 12–15 September 1998).

6. The doctors' dilemma

1. N. Yildirim, *A History of Healthcare in Istanbul* (Istanbul: Istanbul 2010 European Capital of Culture Agency and Istanbul University, 2010), p. 134.
2. Dr Marcou, 'Report on the sanitary situation in Soviet Russia', Correspondance politique et commerciale, série Z Europe, URSS (1918–1940), Cote 117CPCOM (Le centre des archives diplomatiques de la Courneuve, France).
3. H. A. Maureira, '"Los culpables de la miseria": poverty and public health during the Spanish influenza epidemic in Chile, 1918–1920', PhD thesis (Georgetown University, 2012), p. 237.
4. B. J. Andrews, 'Tuberculosis and the Assimilation of Germ Theory in China, 1895–1937', *Journal of the History of Medicine and Allied Sciences*, January 1997; 52:142.
5. D. G. Gillin, *Warlord: Yen Hsi-shan in Shansi Province 1911–1949* (Princeton: Princeton University Press, 1967), p. 36.
6. P. T. Watson, 'Some aspects of medical work', *Fenchow*, October 1919; 1(2):16.
7. N. M. Senger, 'A Chinese Way to Cure an Epidemic', *The Missionary Visitor* (Elgin, IL: Brethren Publishing House, February 1919), p. 50.
8. A. W. Hummel, 'Governor Yen of Shansi', *Fenchow*, October 1919; 1(2):23.

7. The wrath of God

1. R. Collier, *The Plague of the Spanish Lady: October 1918–January 1919* (London: Macmillan, 1974), pp. 30–1.
2. P. Ziegler, *The Black Death* (London: Penguin, 1969), p. 14.
3. In Phillips and Killingray (eds.), 'Introduction', p. 6.
4. A. W. Crosby, p. 47.
5. Letter written to Richard Collier, 16 May 1972. Unpublished. In the collection of the Imperial War Museum, London.
6. Survey published by the Pew Research Center in 2007: http://www.pewresearch.org/daily-number/see-aids-as-gods-punishment-for-immorality/.
7. J. de Marchi, *The True Story of Fátima* (St Paul: Catechetical Guild Educational Society, 1952), http://www.ewtn.com/library/MARY/tsfatima.htm.
8. *Boletín Oficial de la Diócesis de Zamora*, 8 December 1914.
9. J. Baxter, *Buñuel* (London: Fourth Estate, 1995), p. 19.
10. J. G.-F. del Corral, *La epidemia de gripe de 1918 en al provincia de Zamora. Estudio estadistico y social* (Zamora: Instituto de Estudios Zamoranos 'Florián de Ocampo', 1995).
11. *Boletín Oficial del Obispado de Zamora,* 15 November 1918.

PART FOUR: THE SURVIVAL INSTINCT

8. Chalking doors with crosses

1. V. A. Curtis, 'Infection-avoidance behaviour in humans and other animals', *Trends in Immunology*, October 2014; 35(10):457–64.
2. C. Engel, *Wild Health: How Animals Keep Themselves Well and What We Can Learn From Them* (London: Phoenix, 2003), pp. 215–17.
3. F. Gealogo, 'The Philippines in the world of the influenza pandemic of 1918–1919', *Philippine Studies*, June 2009; 57(2):261–92.
4. 'Ce que le docteur Roux de l'Institut Pasteur pense de la grippe', *Le Petit Journal*, 27 October 1918.
5. G. W. Rice, 'Japan and New Zealand in the 1918 influenza pandemic', in Phillips and Killingray (eds.), p. 81.
6. R. Chandavarkar, 'Plague panic and epidemic politics in India, 1896–1914', in Terence Ranger and Paul Slack (eds.), *Epidemics & Ideas:*

Essays on the Historical Perception of Pestilence (Cambridge: Cambridge University Press, 1992), pp. 203–40.

7. Ibid., p. 229. From a report written by an executive health officer in Bombay.
8. N. Tomes, '"Destroyer and teacher": managing the masses during the 1918–1919 influenza pandemic', *Public Health Reports*, 2010; 125(S3):48–62.
9. Ibid.
10. E. Tognotti, 'Lessons from the history of quarantine, from plague to influenza A', *Emerging Infectious Diseases*, February 2013; 19(2):254–9.
11. C. See, 'Alternative menacing', *Washington Post*, 25 February 2005.
12. F. Aimone, 'The 1918 influenza epidemic in New York City: a review of the public health response', *Public Health Reports*, 2010; 125(S3):71–9.
13. A. M. Kraut, 'Immigration, ethnicity, and the pandemic', *Public Health Reports*, 2010;125(S3):123–33.
14. L. M. DeBauche, *Reel Patriotism: The Movies and World War I* (Madison: University of Wisconsin Press, 1997), p. 149.
15. J. Stella, *New York*, translated by Moyra Byrne (undated).
16. A. M. Kraut, *Silent Travelers: Germs, Genes, and the 'Immigrant Menace'* (Baltimore: Johns Hopkins University Press, 1995), p. 125.
17. Excess mortality rates (a measure of the number of people who died over and above what might have been expected in a 'normal' or non-pandemic year) were 40 and 55 per cent higher, respectively, in Boston and Philadelphia than in New York.
18. Olson D.R. et al. 'Epidemiological evidence of an early wave of the 1918 influenza pandemic in New York City, *Proceedings of the National Academy of Sciences* 2005 Aug 2; 102(31):11059–11063.
19. A. M. Kraut, 'Immigration, ethnicity, and the pandemic', *Public Health Reports*, 2010; 125(S3):123–33.
20. R. J. Potter, 'Royal Samuel Copeland, 1868–1938: a physician in politics', PhD thesis (Western Reserve University, 1967).
21. Percy Cox to George N. Curzon, Tehran, 8 March 1920, insert # 1, Anthony R. Neligan to Percy Cox, FO 371/3892 (London: Public Records Office).
22. W. G. Grey, Meshed Diary No. 30, for the week ending 27 July 1918. British Library, London: IOR/L/PS/10/211.

23. M. G. Majd, *The Great Famine and Genocide in Persia, 1917–1919* (Lanham: University Press of America, 2003).
24. The Meshed pilgrimage, P4002/1918, India Office Records (London: British Library).
25. W. Floor, 'Hospitals in Safavid and Qajar Iran: an enquiry into their number, growth and importance', in F. Speziale (ed.), *Hospitals in Iran and India, 1500–1950s* (Leiden: Brill, 2012), p. 83.
26. W. M. Miller, *My Persian Pilgrimage: An Autobiography* (Pasadena: William Carey Library, 1989), p. 56.
27. R. E. Hoffman, 'Pioneering in Meshed, The Holy City of Iran; Saga of a Medical Missionary', ch. 4: 'Meshed, the Holy City' (archives of the Presbyterian Historical Society, Philadelphia, undated manuscript).
28. L. I. Conrad, 'Epidemic disease in early Islamic society', in Ranger and Slack (eds.), pp. 97–9.
29. Document number 105122/3, Documentation Centre, Central Library of Astan Quds Razavi, Mashed.
30. W. M. Miller, p. 61.
31. Hoffman, p. 100.

9. The placebo effect

1. G. Heath and W. A. Colburn, 'An evolution of drug development and clinical pharmacology during the twentieth century', *Journal of Clinical Pharmacology*, 2000; 40:918–29.
2. A. Noymer, D. Carreon and N. Johnson, 'Questioning the salicylates and influenza pandemic mortality hypothesis in 1918–1919', *Clinical Infectious Diseases*, 15 April 2010; 50(8):1203.
3. Nava, p. 202.
4. B. Echeverri, in Phillips and Killingray (eds.), p. 179.
5. Report by Mathis and Spillmann of the 8th Army, Northern Region, 16 October 1918; and 'Une cure autrichienne de la grippe espagnole', memo dated 2 November 1918, Centre de documentation du Musée du Service de santé des armées, Carton 813.
6. P. Lemoine, *Le Mystère du placebo* (Paris: Odile Jacob, 2006).
7. V. A. Kuznetsov, 'Professor Yakov Yulievich Bardakh (1857–1929): pioneer of bacteriological research in Russia and Ukraine', *Journal of Medical Biography*, August 2014; 22(3):136–44.

8. A. Rowley, *Open Letters: Russian Popular Culture and the Picture Postcard 1880–1922* (Toronto: University of Toronto Press, 2013).
9. *Odesskiye Novosti* (*Odessa News*), 2 October 1918.
10. J. Tanny, *City of Rogues and Schnorrers: Russia's Jews and the Myth of Old Odessa* (Bloomington: Indiana University Press, 2011), p. 158.
11. V. Khazan, *Pinhas Rutenberg: From Terrorist to Zionist, Volume I: Russia, the First Emigration (1879-1919)* (in Russian) (Moscow: Мосты култьуры, 2008), p. 113.
12. There is often confusion over dates in Russia in early 1919. The Soviets had imposed the Gregorian calendar in 1918, but in the brief interlude in which the Whites were in power in 1919, they re-imposed the old-style Julian calendar. Dates relating to Vera Kholodnaya's illness and death are given according to the Gregorian calendar. To obtain the Julian equivalent, subtract 13 days.
13. Kuznetsov.

10. Good Samaritans

1. J. Drury, C. Cocking and S. Reicher, 'Everyone for themselves? A comparative study of crowd solidarity among emergency survivors', *British Journal of Social Psychology*, September 2009; 48(3):487–506.
2. D. Defoe, *Journal of the Plague Year* (1722).
3. J. G. Ellison, '"A fierce hunger": tracing impacts of the 1918–19 influenza epidemic in south-west Tanzania', in Phillips and Killingray (eds.), p. 225.
4. S. J. Huber and M. K. Wynia, 'When pestilence prevails . . . physician responsibilities in epidemics', *American Journal of Bioethics*, Winter 2004; 4(1):W5–11.
5. W. C. Williams, *The Autobiography of William Carlos Williams* (New York: Random House, 1951), pp. 159–60.
6. M. Jacobs, *Reflections of a General Practitioner* (London: Johnson, 1965), pp. 81–3.
7. *La Croix-Rouge suisse pendant la mobilisation 1914–1919* (Berne: Imprimerie Coopérative Berne, 1920), pp. 62–3.
8. Dos Santos.
9. S. Caulfield, *In Defense of Honor: Sexual Morality, Modernity, and Nation in Early-Twentieth-Century Brazil* (Durham and London: Duke University Press, 2000), p. 2. See also Dos Santos for commentary.

10. K. Miller, 'Combating the "Flu" at Bristol Bay', *The Link* (Seattle, WA: Alumni Association of Providence Hospital School of Nursing, 1921), pp. 64–66.
11. H. Stuck, *A Winter Circuit of Our Arctic Coast: A Narrative of a Journey with Dog-Sleds Around the Entire Arctic Coast of Alaska* (New York: Charles Scribner's Sons, 1920), p. ix.
12. J. W. VanStone, *The Eskimos of the Nushagak River: An Ethnographic History* (Seattle and London: University of Washington Press, 1967), pp. 3–4.
13. M. Lantis, 'The Religion of the Eskimos', in V. Ferm (ed.), *Forgotten Religions* (New York: The Philosophical Library, 1950), pp. 309–39.
14. H. Napoleon, *Yuuyaraq: The Way of the Human Being* (Fairbanks: Alaska Native Knowledge Network, 1996), p. 5.
15. J. Branson and T. Troll (eds.), *Our Story: Readings from South-west Alaska* (Anchorage: Alaska Natural History Association, 2nd edition, 2006), p. 129.
16. Report of L. H. French to W. T. Lopp, 8 April 1912, Department of the Interior. In Branson and Troll (eds.), p. 124.
17. E. A. Coffin Diary, 1919–1924. Alaska State Library Historical Collections, MS 4-37-17.
18. J. B. McGillycuddy, *McGillycuddy, Agent: A Biography of Dr Valentine T. McGillycuddy* (Stanford: Stanford University Press, 1941), p. 278.
19. Ibid., republished as *Blood on the Moon: Valentine McGillyCuddy and the Sioux* (Lincoln and London: University of Nebraska Press, 1990), p. 285.
20. K. Miller.
21. S. Baker. Warden's Letter to the Commissioner of Fisheries, Bureau of Fisheries, Department of Commerce, Seattle, WA, 26 November 1919. National Archives, Washington DC. Record Group 22: US Fish and Wildlife Service. Also the source of the previous quote.
22. Coffin.
23. A. B. Schwalbe, *Dayspring on the Kuskokwim* (Bethlehem, PA: Moravian Press, 1951), pp. 84–85.
24. Report of D. Hotovitzky to His Eminence Alexander Nemolovsky, Archbishop of the Aleutian Islands and North America, 10 May 1920. Archives of the Orthodox Church in America. Also the source of the following quote.
25. Nushagak was a village across the Nushagak River from Dillingham.

26. Report of C. H. Williams, superintendent, Alaska Packer's Association, in Branson and Troll (eds.), pp. 130–31.
27. VanStone.
28. K. Miller.

PART FIVE: POST MORTEM

11. The hunt for patient zero

1. E. N. LaMotte, *Peking Dust* (New York: The Century Company, 1919), Appendix II.
2. A. Witchard, *England's Yellow Peril: Sinophobia and the Great War* (London: Penguin, China Penguin Special, 2014).
3. Y-l. Wu, *Memories of Dr Wu Lien-Teh, Plague Fighter* (Singapore: World Scientific, 1995), pp. 32–33.
4. L-t. Wu, 'Autobiography', *Manchurian Plague Prevention Service Memorial Volume 1912–1932* (Shanghai: National Quarantine Service, 1934), p. 463.
5. For one exposition of this theory, see: M. Humphries, 'Paths of infection: the First World War and the origins of the 1918 influenza pandemic', *War in History*, 2013; 21(1):55–81.
6. U. Close, *In the Land of the Laughing Buddha: the Adventures of an American Barbarian in China* (New York: G. P. Putnam's Sons, 1924), pp. 39–42. Upton Close was Josef Washington Hall's pen name.
7. J. S. Oxford et al., 'World War I may have allowed the emergence of "Spanish" influenza', *Lancet Infectious Diseases*, February 2002;2:111–14.
8. J. Stallworthy, *Wilfred Owen* (London: Chatto & Windus, 1974).
9. J. A. B. Hammond, W. Rolland and T. H. G. Shore, 'Purulent bronchitis: a study of cases occurring amongst the British troops at a base in France', *Lancet*, 1917; 193:41–4.
10. A. Abrahams et al., 'Purulent bronchitis: its influenza and pneumococcal bacteriology', *Lancet*, 1917; 2:377–80.
11. Personal correspondence with local historian Pierre Baudelicque.
12. Personal correspondence with Douglas Gill.
13. J. M. Barry, 'The site of origin of the 1918 influenza pandemic and its public health implications', *Journal of Translational Medicine*, 2004; 2:3.
14. D. A. Pettit and J. Bailie, *A Cruel Wind: Pandemic Flu in America, 1918–1920* (Murfreesboro: Timberlane Books, 2008), p. 65.

12. Counting the dead

1. Patterson and Pyle, pp. 17–18.
2. 2.5 per cent is the case fatality rate often quoted for the Spanish flu. Note, however, that it doesn't fit with either Patterson and Pyle's or Johnson and Müller's updated death tolls. If one in three people on earth – roughly 500 million human beings – fell ill, and the 2.5 per cent figure is correct, then 'only' 12.5 million people died. On the other hand, if 50 million people died, as per Johnson and Müller's most conservative estimate, then the case fatality rate (global average) was actually closer to 10 per cent.
3. V. M. Zhdanov et al., *The Study of Influenza* (Reports on Public Health and Medical Subjects, Bethesda: National Institutes of Health, 1958).
4. Report of E. Léderrey on the sanitary situation in Ukraine in 1919, Centre des Archives Diplomatiques de la Courneuve: correspondance politique et commerciale, série Z Europe, URSS (1918–1940).
5. W. Iijima, 'Spanish influenza in China, 1918–1920: a preliminary probe', in Phillips and Killingray (eds.), pp. 101–9.
6. Watson.
7. N. P. A. S. Johnson and J. Müller, 'Updating the accounts: global mortality of the 1918–1920 "Spanish" influenza pandemic', *Bulletin of the History of Medicine*, Spring 2002; 76(1):105–15.

PART SIX: SCIENCE REDEEMED

13. *Aenigmoplasma influenzae*

1. R. Dujarric de la Rivière, *Souvenirs* (Périgueux: Pierre Fanlac, 1961), p. 110.
2. Archives de l'Institut Pasteur, fonds Lacassagne (Antoine), Cote LAC.B1.
3. R. Dujarric de la Rivière, 'La grippe est-elle une maladie à virus filtrant?', Académie des sciences (France). Comptes rendus hebdomadaires des séances de l'Académie des sciences. Séance du 21 octobre 1918, pp. 606–7.
4. É. Roux, 'Sur les microbes dits "invisibles"', *Bulletin de l'Institut Pasteur*, 1903(1):7.

14. Beware the barnyard

1. J. van Aken, 'Is it wise to resurrect a deadly virus?', *Heredity*, 2007; 98:1–2.
2. Intriguingly, in 1977, H1N1 was found to have re-emerged in the world. When scientists analysed its genome, they found that it was 'missing' decades of evolution – as if it had been kept in suspended animation somewhere. In fact, though the theory has never been proven, many suspect that a frozen laboratory strain was accidentally released into the general population.
3. R. D. Slemons et al., 'Type-A influenza viruses isolated from wild free-flying ducks in California', *Avian Diseases*, 1974; 18:119–24.
4. C. Hannoun and J. M. Devaux, 'Circulation of influenza viruses in the bay of the Somme River', *Comparative Immunology, Microbiology & Infectious Diseases*, 1980; 3:177–83.
5. For ease, the seasons mentioned in this discussion of the virus's evolution over the course of the pandemic are those of the northern hemisphere.
6. D. S. Chertow et al., 'Influenza circulation in United States Army training camps before and during the 1918 influenza pandemic: clues to early detection of pandemic viral emergence', *Open Forum Infectious Diseases*, Spring 2015; 2(2):1–9.
7. M. A. Beck, J. Handy and O. A. Levander, 'Host nutritional status: the neglected virulence factor', *Trends in Microbiology*, September 2004; 12(9):417–23.
8. P. W. Ewald, 'Transmission modes and the evolution of virulence, with special reference to cholera, influenza, and AIDS', *Human Nature*, 1991; 2(1):1–30.
9. M. Worobey, G.-Z. Hana and A. Rambaut, 'Genesis and pathogenesis of the 1918 pandemic H1N1 influenza A virus', *Proceedings of the National Academy of Sciences*, 3 June 2014; 111(22):8107–12.
10. F. Haalboom, '"Spanish" flu and army horses: what historians and biologists can learn from a history of animals with flu during the 1918–1919 influenza pandemic', *Studium*, 2014; 7(3):124–39.
11. J. K. Taubenberger and D. M. Morens, '1918 influenza: the mother of all pandemics', *Emerging Infectious Diseases*, January 2006; 12(1):15–22.

15. The human factor

1. S.-E. Mamelund, 'A socially neutral disease? Individual social class, household wealth and mortality from Spanish influenza in two socially contrasting parishes in Kristiania 1918–19', *Social Science & Medicine*, February 2006; 62(4):923–40.
2. C. E. A. Winslow and J. F. Rogers, 'Statistics of the 1918 epidemic of influenza in Connecticut', *Journal of Infectious Diseases*, 1920; 26:185–216.
3. C. J. L. Murray et al., 'Estimation of potential global pandemic influenza mortality on the basis of vital registry data from the 1918–20 pandemic: a quantitative analysis', *Lancet*, 2006; 368:2211–18.
4. C. Lim, 'The pandemic of the Spanish influenza in colonial Korea', *Korea Journal*, Winter 2011:59–88.
5. D. Hardiman, 'The influenza epidemic of 1918 and the Adivasis of Western India', *Social History of Medicine*, 2012; 25(3):644–64.
6. P. Zylberman, 'A holocaust in a holocaust: the Great War and the 1918 Spanish influenza epidemic in France', in Phillips and Killingray (eds.), p. 199.
7. V. N. Gamble, '"There wasn't a lot of comforts in those days": African Americans, public health, and the 1918 influenza epidemic', *Public Health Reports*, 2010; 125(S3):114–22.
8. G. D. Shanks, J. Brundage and J. Frean, 'Why did many more diamond miners than gold miners die in South Africa during the 1918 influenza pandemic?', *International Health*, 2010; 2:47–51.
9. M. C. J. Bootsma and N. M. Ferguson, 'The effect of public health measures on the 1918 influenza pandemic in US cities', *Proceedings of the National Academy of Sciences*, 1 May 2007; 104(18):7588–93.
10. A. Afkhami, 'Compromised constitutions: the Iranian experience with the 1918 influenza pandemic', *Bulletin of the History of Medicine*, Summer 2003; 77(2):367–92.
11. A. Noymer, 'The 1918 influenza pandemic hastened the decline of tuberculosis in the United States: an age, period, cohort analysis', *Vaccine*, 22 July 2011; 29(S2):B38–41.
12. C. V. Wiser, 'The Foods of an Indian Village of North India', *Annals of the Missouri Botanical Garden*, November 1955; 42(4):303–412.

13. F. S. Albright et al., 'Evidence for a heritable predisposition to death due to influenza (2008)', *Journal of Infectious Diseases*, 1 January 2008; 197(1):18–24.
14. M. J. Ciancanelli, 'Infectious disease. Life-threatening influenza and impaired interferon amplification in human IRF7 deficiency', *Science*, 24 April 2015; 348(6233):448–53.

PART SEVEN: THE POST-FLU WORLD

16. The green shoots of recovery

1. A. Ebey, 35th annual report for the year ending 29 February 1920, Church of the Brethren, p. 16.
2. S. Chandra, G. Kuljanin and J. Wray, 'Mortality from the influenza pandemic of 1918–1919: the case of India', *Demography*, 2012; 49:857–65.
3. S.-E. Mamelund, 'Can the Spanish Influenza pandemic of 1918 explain the baby-boom of 1920 in neutral Norway?', Memorandum No. 01/2003 (Oslo: Department of Economics, University of Oslo, 2003).
4. For example: H. Lubinski, 'Statistische Betrachtungen zur Grippepandemie in Breslau 1918–22', *Zentralblatt für Bakteriologie, Parasitenkunde und Infektionskrankheiten*, 1923–4; 91:372–83.
5. A. Noymer and M. Garenne, 'The 1918 influenza epidemic's effects on sex differentials in mortality in the United States', *Population and Development Review*, September 2000; 26(3):565–81.
6. J. W. Harris, 'Influenza occurring in pregnant women, a statistical study of thirteen hundred and fifty cases', *Journal of the American Medical Association*, 3 April 1919; 72:978–80.
7. D. Almond, 'Is the 1918 influenza pandemic over? Long-term effects of in utero influenza exposure in the post-1940 US population', *Journal of Political Economy*, 2006; 114(4):672–712.
8. Personal correspondence with Sue Prideaux.
9. K. A. Menninger, 'Influenza and schizophrenia. An analysis of post-influenzal "dementia precox," as of 1918, and five years later further studies of the psychiatric aspects of influenza', *American Journal of Psychiatry*, June 1994; (S6):182–7. 1926.

10. Wellcome Film of the Month: *Acute Encephalitis Lethargica* (1925), http://blog.wellcome.ac.uk/2012/11/02/acute-encephalitis-lethargica-1925/.
11. D. Tappe and D. E. Alquezar-Planas, 'Medical and molecular perspectives into a forgotten epidemic: encephalitis lethargica, viruses, and high-throughput sequencing', *Journal of Clinical Virology*, 2014; 61:189–95.
12. O. Sacks, *Awakenings* (London: Picador, 1983), pp. 105–7.
13. R. R. Edgar and H. Sapire, *African Apocalypse: The Story of Nontetha Nkwenkwe, a Twentieth–Century South African Prophet* (Johannesburg: Witwatersrand University Press, 2000).

17. Alternate histories

1. Ziegler, p. 199.
2. Personal correspondence with Sofie Frackowiak.
3. M. Karlsson, T. Nilsson and S. Pichler, 'The impact of the 1918 Spanish flu epidemic on economic performance in Sweden: an investigation into the consequences of an extraordinary mortality shock', *Journal of Health Economics*, 2014; 36:1–19.
4. E. Brainerd and M. V. Siegler, 'The Economic Effects of the 1918 Influenza Epidemic', Discussion paper no. 3791, February 2003 (London: Centre for Economic Policy Research).
5. S. A. Wurm, 'The language situation and language endangerment in the Greater Pacific area', in M. Janse and S. Tol (eds.), *Language Death and Language Maintenance: Theoretical, Practical and Descriptive Approaches* (Amsterdam: John Benjamins Publishing Company, 2003).
6. G. Kolata, *Flu: The Story of the Great Influenza Pandemic of 1918 and the Search for the Virus That Caused It* (New York: Touchstone, 1999), p. 260.
7. 1994 Alaska Natives Commission report, volume 1, http://www.alaskool.org/resources/anc/anc01.htm#undoing
8. Napoleon, p. 12.

18. Anti-science, science

1. M. Bitsori and E. Galanakis, 'Doctors versus artists: Gustav Klimt's Medicine', *British Medical Journal*, 2002; 325:1506–8.

2. *New York Times*, 17 October 1918.
3. J. C. Whorton, *Nature Cures: The History of Alternative Medicine in America* (Oxford: Oxford University Press, 2002), p. 205.
4. T. Ranger, 'The Influenza Pandemic in Southern Rhodesia: a crisis of comprehension', in *Imperial Medicine and Indigenous Societies* (Manchester: Manchester University Press, 1988).
5. A. Conan Doyle, 'The Evidence for Fairies', *Strand Magazine*, 1921.
6. M. Hurley, 'Phantom Evidence', *CAM*, Easter 2015; 75:31.
7. M. Launay, *Benoît XV (1914–1922): Un pape pour la paix* (Paris: Les Éditions du Cerf, 2014), p. 99.

19. Healthcare for all

1. W. Witte, 'The plague that was not allowed to happen', in Phillips and Killingray (eds.), p. 57.
2. S. G. Solomon, 'The expert and the state in Russian public health: continuities and changes across the revolutionary divide', in D. Porter (ed.), *The History of Public Health and the Modern State* (Amsterdam: Editions Rodopi, 1994).
3. A. A. Afkhami, 'Iran in the age of epidemics: nationalism and the struggle for public health: 1889–1926', PhD thesis (Yale University, 2003), p. 462.
4. M. Micozzi, 'National Health Care: Medicine in Germany, 1918–1945', 1993, https://fee.org/articles/national-health-care-medicine-in-germany-1918-1945/.

20. War and peace

1. E. Jünger, *Storm of Steel*, translated by Michael Hofmann (London: The Folio Society, 2012), p. 239.
2. D. T. Zabecki, *The German 1918 Offensives: A Case Study in The Operational Level of War* (New York: Routledge, 2006).
3. A. T. Price-Smith, *Contagion and Chaos: Disease, Ecology, and National Security in the Era of Globalization* (Cambridge, MA: The MIT Press, 2009).
4. S. Zweig, *The World of Yesterday* (New York: Viking Press, 1943), p. 285.

5. A. A. Allawi, *Faisal I of Iraq* (New Haven: Yale University Press, 2014), p. 223.
6. E. A. Weinstein, 'Woodrow Wilson', in *A medical and psychological biography* (Princeton: Princeton University Press, 1981).
7. Personal correspondence with John Milton Cooper Jr.
8. S. Kotkin, *Stalin, Volume 1: Paradoxes of Power, 1878–1928* (London: Allen Lane, 2014).
9. Davis.
10. M. Echenberg, '"The dog that did not bark": memory and the 1918 influenza epidemic in Senegal', in Phillips and Killingray (eds.), p. 234.
11. M. K. Gandhi, *Autobiography: The Story of My Experiments with Truth* (CreateSpace Independent Publishing Platform, 2012), p. 379.
12. A. Ebey, 35th annual report for the year ending 29 February 1920, Church of the Brethren, p. 17.
13. A. Bhatt, 'Caste and political mobilisation in a Gujarat district', in R. Kothari (ed.), *Caste in Indian Politics* (New Delhi: Orient Longman, 1971), p. 321.
14. A. J. P. Taylor, *English History 1914–1945* (Oxford: Oxford University Press, 1965), pp. 152–3.
15. Letter from Tagore to a friend, 11 May 1919, *Young India*, August 1919, volume 2.

21. Melancholy muse

1. W. L. Phelps, 'Eugene O'Neill, Dramatist', *New York Times*, 19 June 1921.
2. F. B. Smith, 'The Russian Influenza in the United Kingdom, 1889–1894', *Social History of Medicine*, 1995; 8(1):55–73.
3. J. Iwaszkiewicz, 'The History of "King Roger"', *Muzyka*, 1926, number 6, http://drwilliamhughes.blogspot.fr/2012/05/jarosaw-iwaszkiewicz-history-of-king.html.
4. P. Gay, *Freud: A Life for our Time* (New York: W. W. Norton & Company, 2006), p. 392.
5. R. Stach, p. 262.
6. Davis, p. 109.
7. L. M. Bertucci, *Influenza, a medicina enferma: ciência e prácticas de cura na época da gripe espanhola em São Paulo* (Campinas: UNICAMP, 2004), p. 127.

8. A. Montague, 'Contagious Identities: literary responses to the sanitarist and eugenics movement in Brazil', PhD thesis (Brown University, 2007).
9. S.Wang, *Lu Xun: A Biography* (Beijing: Foreign Languages Press, 1984), pp. 27–9.
10. Andrews, pp. 141–2.
11. S. T. Nirala, *A Life Misspent*, translated by Satti Khanna (Noida, UP: HarperCollins, 2016), pp. 53–4.

PART EIGHT: ROSCOE'S LEGACY

1. D. A. Kirby, *Lab Coats in Hollywood: Science, Scientists, and Cinema* (Cambridge, MA: The MIT Press, 2010), location 1890 (Kindle version).
2. A. Gulland, 'World invests too little and is underprepared for disease outbreaks, report warns', *British Medical Journal*, 2016; 352:i225.
3. J. Shaman and M. Lipsitch, 'The El Niño–Southern Oscillation (ENSO)–pandemic influenza connection: coincident or causal?', *Proceedings of the National Academy of Sciences*, 26 February 2013; 110(S1):3689–91.
4. Audubon, *Birds and Climate Change Report*, 2014, http://climate.audubon.org.
5. N. A. Christakis and J. H. Fowler, 'Social network sensors for early detection of contagious outbreaks', *PLOS One*, 15 September 2010; 5(9):e12948.
6. R. P. P. Almeida, 'Can Apulia's olive trees be saved?', *Science*, 22 July 2016; 353:346–8.

AFTERWORD: ON MEMORY

1. H. Phillips, 'The recent wave of 'Spanish' flu historiography', *Social History of Medicine*, 2014. doi:10.1093/shm/hku066.
2. J. W. Thompson, 'The aftermath of the Black Death and the aftermath of the Great War', *American Journal of Sociology*, 1921; 26(5):565–72.
3. G. D. Shanks, 'Legacy of the 1914–18 war 1: How World War 1 changed global attitudes to war and infectious diseases', *Lancet*, 2014; 384:1699–707.

4. H. L. Roediger and M. Abel, 'Collective memory: a new arena of cognitive study', *Trends in Cognitive Sciences*, 2015; 19(7):359–61.
5. http://numismatics.free.fr/FIM/FIM%20-%20Medaille%20des%20EpidemiesV3.0.pdf.
6. D. Gill, 'No compromise with truth: Vera Brittain in 1917', *War and Literature*, Yearbook V, 1999:67–93.
7. M. Forrier, *Edmond Rostand dans la Grande Guerre 1914–1918* (Orthez, France: Editions Gascogne, 2014), p. 414.

Index